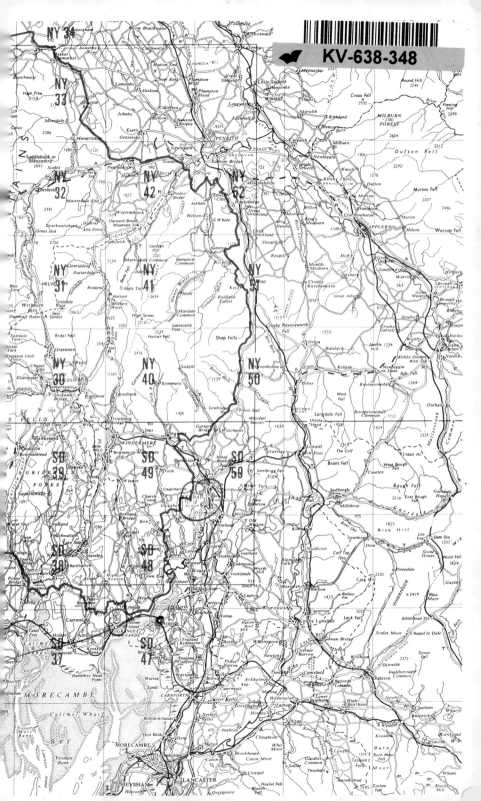

LAKE DISTRICT

NATIONAL PARK

Stonethwaite Beck, Borrowdale, looking upstream towards Eagle Crag

LAKE DISTRICT

NATIONAL PARK GUIDE NO. 6

EDITED BY
THE COUNTRYSIDE COMMISSION

ISSUED FOR
THE COUNTRYSIDE COMMISSION

LONDON
HER MAJESTY'S STATIONERY OFFICE

© *Crown copyright 1975*
First published 1969
Second edition 1975

ISBN 0 11 700482 0

PREFACE

THE first edition of this guide appeared in 1969 and I am very pleased to welcome this second edition, which shows extensive revision and up-dating of the previous material. The fact that the first edition has so quickly gone out of print is praise enough for the various contributors, and I am sure that this second edition will enjoy similar popularity.

It is sad that three of the contributors to the first edition are no longer with us to revise their previous articles. In fact, as a result of the inevitable delays involved in the process of publication, all three had died before the previous edition was available to the public, and I would like to record here a brief tribute to all three.

Professor Pearsall, F.R.S., was the editor of the first edition. He was a 'local lad' and like Lord Birkett was educated at Ulverston Grammar School. After leaving the University of Manchester he held posts successively at Leeds and Sheffield Universities and then until his retirement was Quain Professor of Botany in University College, London. A founder member of the Freshwater Biological Association, he acted as honorary director between 1931 and 1937 and was chairman of the council from 1954 onwards. Few have done as much as he did to bring the association from its humble birth in a period of economic depression to its present position as one of the most famous bodies of its kind in the world.

The other two were Kenneth Dobell and Philip Cleave, and I knew them both very well, as they were original members of the Lake District Planning Board when it was created in 1951.

Kenneth Dobell was the first vice-chairman of the Board, chairman from 1957 to 1964, and a member until a few weeks before his death in 1968. He was deeply interested in the protection and development of the National Park and gave much time and thought to its many problems—and to possible solutions to the problems.

Philip Cleave's knowledge of the Lake District was quite incredible. He seemed to know every farm and stream, and this was all the more remarkable because he never owned a car and had to rely on his legs, public transport, or the odd lift from a friend. I never knew him to be late for a meeting or site inspection. He was not a 'local lad' but had made the Lake District his love and his home.

We all owe a great deal to these three men for all they did towards the establishment of the Lake District National Park.

R. G. GRICE
Chairman
Lake District Special Planning Board

CONTENTS

	page
PREFACE R. G. GRICE, Chairman of the Lake District Planning Board	iii
1. LOOKING AT THE LAKES Norman Nicholson	1
2. ROCKS AND SCENERY Dr A. G. Lunn, *Lecturer, Department of Adult Education, University of Newcastle upon Tyne*	7
3. VEGETATION AND PLANTS Dr R. G. H. Bunce, *Senior Scientific Officer, Nature Conservancy Council, Merlewood Research Station*	17
4. ANIMALS OF THE NATIONAL PARK Dr R. J. Elliott, *Nature Conservancy Council Conservation Officer, Wales*	36
5. REMAINS LEFT BY ANCIENT PEOPLES Miss Clare I. Fell, *Cumberland and Westmorland Antiquarian and Archaeological Society*	52
6. THE MEDIEVAL SCENE AND AFTER Canon J. C. Dickinson, *Lecturer in Theology, University of Birmingham*	59
7. SOCIAL AND ECONOMIC CHANGE G. P. Jones, *formerly Professor of Economics, University of Sheffield*	64
8. INDUSTRIAL ARCHAEOLOGY Dr J. D. Marshall, *Reader in North-West Regional History, University of Lancaster*, and M. Davies-Shiel, *Geography and Environmental Studies Master, Lakes School*	72

Contents

9. HOW TO ENJOY THE NATIONAL PARK
   A. H. Griffin 79

10. ANCIENT WAYS AND MODERN WALKS
    Roland Wade 88

11. FELL WALKING AND ROCK CLIMBING
    J. Robert Files 102

12. TRADITIONAL SPORTS
    A. H. Griffin 114

13. LITERARY ASSOCIATIONS
    George Bott 122

14. WEATHER AND CLIMATE OF THE LAKE COUNTIES
    Professor Gordon Manley, *formerly Professor of Geography, University of Lancaster* ... 129

SHORT BIBLIOGRAPHY 139

APPENDICES
  I. Scheduled Ancient Monuments 141
  II. Some Useful Addresses 145

INDEX 153

ILLUSTRATIONS

PLATES

Stonethwaite Beck, Borrowdale	*frontispiece*
	facing page
I. Derwent Water from Latrigg	22
II. Coniston Fells, Langdale, and Borrowdale	
III. Looking across Coniston Water	23
IV. Red deer grazing in Grizedale Forest	
V. Castlerigg stone circle	38
VI. Sheep farming	
VII. Windermere Hotel	39
VIII. Comb Gill corn and saw mill, Borrowdale	
IX. Brockhole National Park Centre	86
X. Ancient ways and modern walkers	
XI. Fell walking	87
XII. Cumberland wrestling	102
XIII. Brantwood, the home of John Ruskin	
XIV. Cumulus cloud developing over High Street	103

FIGURES

1. Ordnance Survey 2½-inch map sheet numbers for the National Park area	*inside front cover page*
2. Geological map	following page 6
3. Minerals	9
4. Plants of the lower fells	30
5. Butterflies and moths	40
6. Invertebrates living in stony streams	46

7. Large stone circles in Cumbria
   Packhorse bridges
   Scandinavian remains
   Bronze Age burials
   　　　　　　　　　　　　　　page 54

   　　　　　　　　　　　　　　following page
8. Walkers' and climbers' map　118

ACKNOWLEDGMENTS

Thanks are due to all those who helped with the preparation of this guide, and particularly to the authors of the various chapters. The material for Figure 8 was prepared by J. Robert Files and the scheduled list of Ancient Monuments was contributed by the Department of the Environment.

The Commission are also indebted to the photographers named below whose work is reproduced:

Geoffrey Berry for Plates I, III, IV, X, XI, and XIV; Brantwood Education Trust for Plate XIII; M. Davies-Shiel for Plate VIII; Forestry Commission for Plate IV; Lake District Planning Board for Plates VI, VII, and IX; Tom Parker for the frontispiece; *Westmorland Gazette* for Plate XII; C. H. Wood, Bradford, for Plate II.

The cover, geological map, and all the drawings were designed and produced by the Lake District Planning Board.

1
Looking at the Lakes

by NORMAN NICHOLSON

THERE is a look of permanence about mountains. 'As old as the hills' is one of our commonest clichés, and when a Cumberland exile thinks of home it is always the hills that come into his mind as symbols of what will not have changed. They say, in south Cumberland, that those born within sight of Black Combe will return to it from whatever part of the world they may be dispersed to, and the same is said, no doubt, of Sca Fell, Coniston Old Man, Helvellyn and Saddleback (Blencathra). We take it for granted that much of the countryside will alter from year to year, as new houses are built, new roads made, and new industries introduced, But the hills, we feel, will always stay the same.

In this, of course, we are quite wrong, for the hills themselves are the product of change. The oldest rocks in the Lakes (those around Skiddaw) were formed from the erosion by rain and rivers of the remains of still older rocks. The crags of the central dales come from ashes and lava thrown out by ancient volcanoes. It was change, too, the catastrophic change of Ice Age glaciation, which brought to the dales their characteristic shape and gave rise to so much that we regard as typical of dale scenery—lakes, gills, waterfalls, and the steep rake-ends of the fellsides. And though fire and ice may (perhaps only temporarily) be out of action, yet the work of erosion goes on. Every beck is steadily sandpapering the fells and swilling the dust towards the Irish Sea. All the lakes are diminishing in size: the largest, which once stretched from the head of what we call Derwent Water to the foot of Bassenthwaite, being now bisected into two, while scores of smaller ones are already silted up and have disappeared. The process may be too slow to make much observable difference in a lifetime, but, once it is recognised to be taking place, the landscape gathers a new interest. It is no longer merely static. It is seen to be changing, moving, and, in a way, to be alive.

Yet it is not these physical changes—geological or botanical—that I want to write about now, but about the way in which the hills seem to change because of changes in the people who look at them. For what we see in a landscape depends on what we are, on the age we belong to, and on the way we are trained to use our eyes. To the man of the Middle Ages, a man of the eighteenth century, and a man of today, Skiddaw presented three quite different objects, one for each.

To the man of the Middle Ages, the mountains belonged to a land beyond common human experience—a land of rocks, ice, forests, and wild seas, inhabited by monsters and legends. When medieval man looked at mountains he did not see a view: he saw the gates of Hell or the gates of Heaven; he saw the physical shapes of all that was mysterious and terrifying in his own mind. Nor did he see a view when he looked at the more homely, less alarming side of nature which was all around him. For he was the product of a civilisation essentially rural. City and village depended equally on the land. The ploughing of the soil, the growing of crops, the vicissitudes of weather, the almanac of the seasons, were not just a part of his existence—he was a part of theirs. He could not separate himself from the world of birds, beasts, flowers, grass, water, and rock. He could not look at that world as a *view* because he himself was part of the view.

By the eighteenth century, this was beginning to change. Wild nature was no longer very wild. Swamps were being drained; roads were being laid down. At the same time, there was emerging a new class of men: enterprising, inventive, often unscrupulous, always intensely practical—the men who were to make industrial England. Thomas Pennant, for instance, showed enormous interest in the Whitehaven mines, which he visited in 1769, though he was not blind to natural beauty and considered himself a connoisseur of landscape.

Moreover, this new urban civilisation produced more than merely the practical-minded man. It produced also men who were able to look at the countryside with a new detachment, who were able to stand back and enjoy it as a view. 'Picturesque Beauty'—the term which became fashionable towards the end of the eighteenth century —meant, at the time, precisely what it says: 'Beauty which would be effective in a picture'. And it was from pictures, especially from the work of French and Italian landscape painters, that the traveller in search of Picturesque Beauty learnt what to look for. Dr John Brown in his famous *Description of the Lake and Vale of Keswick*, published in 1769, was the first to compare the Lakeland scene with the three artists who were to set the standards for the new taste. At Derwent Water, he says, there are found the three constituent principles of perfection in landscape, 'beauty, horror and immensity'. And to describe these adequately he would need the united powers of Claude Lorraine, Salvator Rosa, and Gaspar Poussin: 'The first should throw his delicate sunshine over the cultivated vales, the scattered cots, the groves, the lakes, and wooded islands. The second should dash out the horror of the rugged cliffs, the steeps, the hanging woods, and foaming water-falls; while the grand pencil of Poussin should crown the whole with the majesty of the impending mountains.'

It was, however, a Cumberland man who did more than anybody else in England to establish the fashion for 'Picturesque Beauty'. William Gilpin, a member of a famous Border family, was born in 1724 at Scaleby Castle near Carlisle, his father later having command of the garrison in that city. He was educated at St. Bees and Oxford, and afterwards took charge of the Rev. Daniel Saxby's school at Cheam—the ancestor of the preparatory school attended by the Prince of Wales. The school seems to have been as successful then as it is now, and by the time he was fifty-three, Gilpin had saved £10,000 and was able to retire from teaching and take a living in the New Forest. He then began to publish journals of a series of summer tours, to the Wye Valley, the South Coast, the Lakes, Wales, and Scotland, which were to revolutionise the Englishman's idea of a holiday. For Gilpin made himself the travelling salesman of the Picturesque, and went about adjudicating the landscape, giving Buttermere or Loch Lomond so many out of ten. He concerned himself solely with the *visual* aspect of the scene, allowing nothing else to affect his judgment. The true character of a district—what we would now call the *feel* of it—interested him not at all. Nor did the life of the people. If men were included in his landscapes it was merely as picturesque objects to suggest the size and proportion of the rest of the scene.

We can get a good idea of the way his mind worked from what he had to say about the landscape which had aroused John Brown to the cry of 'beauty, horror and immensity'. Derwent Water, says Gilpin: 'unquestionably . . . is, in many places, very sweetly romantic . . . but to give it pre-eminence may be paying it perhaps much too high a compliment . . . . In the first place, its form, which in appearance is circular, is less interesting, I think, than the winding sweep of Windermere, and some other lakes . . . . [And] to the formality of its shores may be added the formality of its islands. They are round, regular, and similar spots, as they appear from most points of view; formal in their situation, as well as in their shape; and of little advantage to the scene . . . . But among the greatest objections to this lake is the abrupt, and broken line in several of the mountains, which compose its skreens . . . . We have little of the easy sweep of a mountain-line: at least the eye is hurt with too many tops of mountains, which injure the ideas of simplicity, and grandeur.' (1786.)

We may feel that Gilpin's descriptions, though expressed in the terms of Six Easy Lessons for amateur sketchers, are clearer and more definite than John Brown's. Gilpin was, at least, looking at the scene and not just going into raptures about it. But the man whose eye was so easily hurt by too many tops of mountains was not the man to respect precise detail.

This can be seen from the oval, terra-cotta-coloured aquatints with which he illustrated his books. One cannot call them 'views'. They are only tastefully-romantic generalisations of nowhere in particular. In their own vague way they have something of the charm of a foggy autumnal sunset, but as introductions to the Lakeland scene they are both preposterous and misleading. Yet it was through these pictures, as through coloured spectacles, that the men of taste first learned to look at the Lakes.

Another man who had enormous influence on the taste of the Lakes tourists was Thomas Gray, the poet of the *Elegy*. Gray carried a convex mirror which, he tells us, played its part divinely 'in lower Borrowdale'. But, when he wanted to, he could put down the glass and really look about him, and, unlike Gilpin, he did not restrict his enjoyment to the *visual* aspects of the scene. Instead, he brought an immensely scholarly mind trained to enjoy the landscape in many different ways. He was a botanist and a student of the weather. He was an archaeologist and historian with a special knowledge of heraldry and medieval architecture. He was a philologist with a vast interest not only in the classical languages, but also in Provençal, Old Norse, Icelandic and Anglo-Saxon. He was in fact superbly equipped to appreciate everything that appealed to the eighteenth-century sense of the 'romantic'—wild rocks, dark chasms, mysterious ruins, stone circles on the moors, old churches in the dales. (See also p. 122.)

By the side of Gray's highly sensational impression of Lake scenery, Gilpin's more sedate view seems academic and even 'purist'. For a while, however, Gilpin led the fashion—the young Wordsworth and his brothers were greatly influenced by him. But before long the tourists began to follow Gray and search for the exciting, the eccentric, the Gothic, and the quaint. For them, the crossing of the sands of Morecambe Bay made a spectacular and adventurous prelude to the Lakes tour. They were curious, also, about strange natural phenomena, such as the Helm Wind on Cross Fell, or the great flood at St. John's-in-the-Vale in 1749. They revelled in stories of ghosts and hauntings, such as that of the celebrated phantoms of Souter Fell which seemed to anticipate the Rebellion of the '45 ten years before it took place.

Even some of the inhabitants of the district were infected with the craze for the Gothic. Thomas Wilkinson, a Quaker, indulged in the fancy—surely strange for one of his persuasion—of building a grotto at Yanwath on the banks of the river Eamont. Colonel Braddyll, of

Conishead near Ulverston, constructed a hermitage in his grounds, and employed a full-time hermit, who held the post for twenty years and never had his hair cut. Mr Pocklington turned the largest island of Derwent Water into a real fairground of the Picturesque, with a fort and battery, a mock church, and a miniature Druid's Circle modelled on Castlerigg.

It is easy, of course, to laugh at the foibles of the early tourists, but it would be a mistake to think that they were merely fools. Many of them—like Gray and Gilpin—were men of intellect, education and learning. If their reactions seem to us absurdly exaggerated, that should warn us that our own reactions are just as likely to be coloured and distorted by the fashions of our own times.

And, what is more, those fashions may not be quite so different from the fashions of the late eighteenth century as we usually think. Few people today share the taste for Gothic horrors and sham ruins, but many have a mental picture of the countryside that is quite as unrealistic as Gray's. The townsman often regards the country as quaint, old-world, and charmingly out of date. He still tends to think of country people as simple-minded gaffers, speaking strange dialects, and having uncouth ideas of humour and funds of homely wisdom. And this romantic self-deception—which combines a mild envy with a pleasant sense of urban superiority—is supported on every hand by the propaganda of popular radio and television shows, Christmas cards and calendars, chocolate boxes and embroidered tea-cosies. It is at its best a harmless make-believe, and at its worst a misunderstanding which makes the townsman blind to the realities of the countryside and the rights and problems of the people who live there.

Gilpin's search for Picturesque Beauty may seem still further removed from modern trends, but, in fact, the cult of the view has never been more popular than it is today. Only we take our idea of a view not from Gilpin's vague aquatints but from the photograph. Few of us realise how enormously the photograph has influenced the way we look at a landscape. We have learned from it to watch out for the bold outline, the strong contrast between sunlight and shadow, the abrupt break between foreground and background. And, staring through the windscreens of cars, that is precisely what many people see—a photograph drawn large. Yet, often, they must feel a certain disappointment. Some of the most famous views in the Lakes—Friar's Crag on Derwent Water, for instance, or Wast Water from near the foot of the lake—are so like what the photograph has led us to expect that to see them in reality seems almost an anticlimax. The photograph has done its work so well that the best that the view can do is merely to imitate. Of course, artists can often open our eyes to new features of the landscape that we may miss in a photograph. Gilpin, in his naïve way, made men see beauties they had ignored before. The French Impressionists gave us a new perception of light and colour, and the best-known of the present-day Lake artists make us more sharply aware of the geometrical structure of the fells. But when artists or photographers begin to dictate to our eyes and to substitute their vision for what should be our own, then we are in danger of falling into the error of the Picturesque all over again. So that it may be useful to try to see the Lakes as they once appeared to those first tourists of two hundred years ago. Fashions in landscape are nearly as changeable as fashions in clothes, and to the man of the twenty-second century our photographs and guide-books will look as odd and fanciful as Gilpin and Gray look to us.

To see the Lakes clearly, however, we must pass beyond the photograph or picture; we must penetrate to the living landscape behind the view. We must get out of our cars, feel the rock under our feet, breathe the Cumbrian air, and learn to know something, at least, of the complex organic life of grass, herb and tree, something of the changing pattern of weather, water and rock, and something of the way man has helped to shape the landscape in the past and is shaping it today. In fact, if we really want to look at the world around us, we must learn to use our knowledge, curiosity, reason, imagination, mind and heart and all five senses as well as just our eyesight.

2
Rocks and Scenery

by A. G. LUNN

ON 23 May 1842, Adam Sedgwick, the Reverend Professor of Geology at Cambridge, wrote to William Wordsworth the first of five letters on Lake District geology which were to appear in the several editions of Wordsworth's *Guide to the Lakes*. Sedgwick was adressing 'any intelligent traveller whose senses are open to the beauty of the country around him' and would 'endeavour to avoid technical language' as far as he was able. Wordsworth understood that the interpretation of scenery properly included an account of the geology.

Rock exposures indeed impinge much on the consciousness of visitors to the Lake District, and a great variety of rock types are present. *Igneous* rocks are those which have solidified from molten magma, which was either intruded in bulky masses beneath a cover of already existing rocks or extruded at the surface from volcanic vents as lava or showers of ash. *Granite* is a rock of the first, intrusive, type. It is often pink, is speckled owing to the large mineral crystals of which it is composed; and was quarried formerly for building stone but is used now mainly for road-metal. Its locations are shown on the geological map (Figure 2), as is the area, constituting the rugged central part of the National Park, where extrusive igneous rocks of the Borrowdale Volcanic Group outcrop at the surface (geologists mean the surface that would be there if the loose, unconsolidated deposits of clay, etc., often lying on top were removed).

Unlike granite, the volcanic rocks occur in layers—they are stratified. A sheet of *lava* flowing widely from a vent would be covered by a deposit of ash, which was in turn perhaps covered by another lava flow and so on until a thick pile of strata, fairly horizontal except near the vent itself, accumulated. The mountains—Helvellyn, Sca Fell, Great Gable—are not to be envisaged as volcanoes themselves, but as remnants of the pile of volcanic strata; they have been isolated from one another by valleys eroded much later by rivers and glaciers into the stratified rock sequence. The volcanic ashes, hardened into solid rock, are called *tuffs*. Many of these are conspicuously laminated and the thin layers of an inch or less are picked out on rock faces by weathering. Many more are streaky in appearance because they are composed of small discs, an inch or so in diameter, welded together. They originated in horrendous, incandescent clouds of lava droplets which moved out at high speed from the vents; after landing the droplets were compressed into discs and built up into these *welded-tuffs* or *ignimbrites*. Volcanic rocks composed of large angular stones ejected from volcanoes are *agglomerates*, but can be difficult to distinguish from *flow-breccias*, which were formed when the surfaces of molten lava congealed into crusts only for these to be broken into fragments and reincorporated on further flow. These volcanic rocks have a variety of chemical compositions: they vary in their mix of mineral crystals, but many can be described as *andesites*, which are normally dark grey, green or purple in colour, or *rhyolites*, which are paler. The lava which makes Pavey Ark is an andesite, whilst that of the western crags of Scafell Pike is a rhyolite. Colours, however, can mislead unless a freshly broken-off specimen is examined; surfaces are altered by weathering and by coatings of lichen.

Sedimentary rocks were deposited, usually in the sea, as particles—later to be cemented together—derived from the denudation of rocks elsewhere. Because they were laid down layer by layer they too are stratified, and can be classified according to the sizes of their particles. *Sandstones* are composed of grains of sand size, *siltstones* of finer particles, and rocks which are composed of minute crystals of clay and other minerals are either *mudstone* or, if laminated and fissile, *shale*. One of the most abundant Lake District rocks is none of these but a *greywacke* (pronounced with a final 'ee'), a tough dark-coloured sandstone with a clay matrix. The areas indicated as Skiddaw Slate and Silurian on the map consist of sedimentary strata of shale, mudstone, siltstone and greywacke, with some sandstone, alternating one above the other. *Limestone*, which outcrops in the Carboniferous Limestone ring around the high fells, consists of the calcium carbonate skeletal remains of marine organisms.

The last great group of rocks is *metamorphic*, formed by the alteration of rocks already present. *Slate*, so widespread in the Lake District, is a weakly metamorphosed rock. It possesses cleavage: parallel planes, unrelated to the original bedding, along which the rock may be split into thin sheets. Cleavage is induced by several mechanisms. Compressive stress and associated chemical change cause the orientation of flaky mineral particles at right angles to the stress; or the forced expulsion upwards of water from unconsolidated sediments reorientates particles. In either condition the planes in which the flaky particles come to lie direct cleavage. A third type arises from very closely spaced fracture-planes. The bedding may be

FIGURE 3 # Minerals

Barytes crystals

Hematite – Kidney iron ore

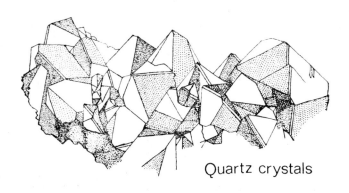

Quartz crystals

3 common types of calcite crystal

Nail head   Dog tooth   Prismatic

at any angle to the cleavage and is often evident, as delicate laminae, in the faces of slates themselves. In the Lake District shales, mudstones, and some volcanic tuffs have been altered to slate, and it is from tuffs that the famous green roofing slates are produced at Honister, Coniston and elsewhere, although now more of the rock is split and sawn into slabs for cladding and general building purposes than into roofing slates. The rough building stone used in much of the region consists in part of the discards of slate quarries but also of non-cleaved lavas and tuffs, as well as dour-coloured greywackes—whatever was locally to hand. Until slate is riven by quarrymen the cleavage represents only potential planes of division, but slate rock has split into thin fragments from natural causes also—stress in the earth's crust after the slate was formed, or exposure to the weather—so that abundant slate scree occurs in the Borrowdale Volcanic area.

Rocks are also classified according to the period at which they were deposited, and conventional names are applied to these divisions of geological time. Fossil content is the basis of division but gives only a relative time-scale which is now supplemented by absolute radiometric dates. The key of the geological map assigns the local rocks to their periods and gives approximate dates. It is a matter of sorrow to some conscientious geologists that no rocks younger than about 200 million years old are known to survive in or around the Lake District (apart from recent glacial and other deposits). During the latter part of this 200-million-year gap, after the area emerged from under the sea for the last time, the topography assumed its present shape as the rocks were weathered and eroded, but of this—owing to the absence of the rocks which could provide evidence of what was going on—nothing is known and only a little may be guessed.

Although the detailed disposition and pattern of outcrop of Lake District rocks is overwhelmingly complex, the main groups into which they are divided on the map show a quite simple geographical pattern. The Skiddaw Slate Group outcrop in an ENE–WSW belt through the northern part and are the oldest rocks present. They consist of many thousands of feet of greywacke, siltstone and mudstone beds, crumpled by stress in the earth's crust into tight folds whose wave-length varies from miles to a few feet and less. The finer-grained rocks of the group are generally cleaved, but rarely yield usable slates. They are flanked to north and south, in similar ENE–WSW belts, by the next younger strata, the Borrowdale Volcanic Group whose lavas, tuffs and slates have already been described. It is the wide southern belt of these rocks which forms the central Lake District, between Keswick and Ambleside. Unlike the Skiddaw Slate strata, the tougher volcanic beds have been thrown only into wide folds, but like them have been intensively faulted such that beds on one side of a fracture-plane moved up, down or sideways in relation to those on the other. Great friction could be generated, and strong rocks were crumbled along the fracture-planes into shatter-zones a few yards wide. Younger again and outcropping south of the main belt of Borrowdale Volcanics is the Coniston Limestone Group, collectively thin and occupying a narrow zone, but whose impure limestones carry calcium-rich soils and lime-loving plants right across the Lake District. Farthest south of the broad ENE-WSW belts is a thick group resembling the Skiddaw Slate which has no other name than that of its period, the Silurian. Here are some workable slates, notably at the huge Burlington quarries at Kirkby-in-Furness, just outside the National Park.

The geological section from north to south across the area shows the general disposition of these old Ordovician and Silurian rocks. They are glimpsed as it were through a window of overlying later rocks, and the reader should compare this section with the map to understand the structural basis of their distribution. As a result of the denudation of the top of a great arched fold trending ENE-WSW, the Skiddaw Slate came to be exposed at its core, with the Borrowdale Volcanics and younger rocks in similar ENE-WSW belts on its flanks. Individual strata within each group also normally trend ENE at outcrop, but the structure is so complex that there are many exceptions. The younger covering rocks, Carboniferous Limestone and New Red Sandstone, drape the central Lake District (as the map and section show) and outcrop on lower ground in more or less concentric belts, quite different in trend from the ENE-WSW belts of old rocks. The Carboniferous Limestone Series consists largely of massive beds of white limestone; the New Red Sandstone of red sandstones and shales.

Each group has produced a characteristic type of scenery; indeed, the contrasts are striking and are a main theme throughout this guide. The mountains of the Skiddaw Slate are typically smooth-sided, vegetated, and conical in shape, whereas those of the Borrowdale Volcanic Group are frequently irregular, stepped, ribbed, rugged, sometimes with precipitous crags and with abundant outcrops of solid rock. Here the exciting mountain scenery of the central Lake District is to be found. From most viewpoints in the northern part the contrast is plain to see. The Skiddaw Slate mountains have smooth slopes because the rocks are relatively homogeneous with respect to their ability to resist erosion, for example by glacier ice, and are also rather weak. Their weakness arises partly from the inherent nature of some shales and mudstones but also from the close-set jointing (natural cracks produced by stress) and cleavage of these, and of intrinsically stronger rocks such as greywackes, which cause them readily to break down into chips and splinters. Their homogeneity is also explained by this weakening of otherwise tough greywackes, which moreover are often thin. The smoothness of the mountain slopes is therefore due to the scarcity of tough strata able to resist erosion and form topographic features, and to the readiness with which the rocks weather down into rubble which moves down slope to smother most small crags that might have been present. Some massive greywackes, however, form crags.

The Borrowdale Volcanic rocks differ in that they include many thick, tough strata, both lavas and some tuffs, as well as weaker ones such as slates. This heterogeneity produces irregular slopes, with the tough beds, up to 100 feet thick and with widely-separated joints, accounting for crags, benches, ribs, plateaux and innumerable rocky knolls. Where joint-bounded blocks are shifted, for example by glacier ice (the blocks later to be dumped as huge boulders), the faces of their sockets form sheer rock faces, ledges, buttresses and chimneys. A further cause of irregularity is that faults shift the tough feature-making strata about, so that a long horizontal crag comes to an abrupt end, only to reappear and continue several hundred feet higher up the slope. Moreover shatter-zones, where they transect tough strata, are lines of weakness picked out as gills, gullies and skyline notches.

The Silurian rocks form lower plateau and valley country. Massive greywackes and sandstone make the more rugged ridges, plateaux and knolls. The granites are uniform in durability and form irregular undulating hills; the Carboniferous limestones form conspicuous desk-shaped uplands with grey-white crags or scars on the fringes of the National Park; and in the New Red Sandstone area, mainly outside the Park entirely, ridges of red sandstone rise through the encumbering glacial deposits. The rocks of each belt are reflected in local building material and field and fell walls.

The radial pattern of the valleys which contain the lakes has for long been explained by assuming that the rivers which eroded the valleys—long before the Ice Ages—flowed off a dome-shaped cover of young rocks, now vanished, so that the pattern was able to become impressed on to the old rocks beneath, whose ENE–WSW trend of outcrop would normally have guided drainage in those directions. This is sheerest speculation; an alternative view is that glaciers were the cause, carving at least some new troughs as they flowed outwards from a centre of snow and ice accumulation in the Sca Fell–Great Gable–Glaramara knot of high mountains. Shatter-zones too would have guided erosion of valleys. Whichever view is correct, it was down these radial valleys that valley glaciers flowed during the most recent glaciation.

Incandescent tuff flows notwithstanding, the Lake District would offer little sensation today were it not for temperatures at the other extreme which prevailed recurrently during the Pleistocene epoch, between about 2 million and 10,000 years ago. It was glacier ice that eroded the basins in which the eponymous lakes lie, and without its effects there would be neither tarns, nor crags, nor screes but rather a subdued, rounded, tame countryside. It is not known how many separate glaciations affected the Lake District; but the area was almost or completely covered by an ice-sheet of the last, which, nourished by snows in this and other upland areas, flowed as far south as the West Midlands and the Wash as recently as 18,000 years ago. This ice had melted away by 12,000 years ago, whether steadily or with occasional lesser readvances is still uncertain, and there was a late, very cold episode between about 10,800 and 10,300 years ago, when ice for the last time, at least so far, occupied the corries and the heads of the valleys.

Indeed during each glaciation there were periods when glacier ice was confined to the high mountain basins or corries; others when ice-streams flowed from the corries and off plateau gathering grounds to coalesce as valley glaciers in the pre-existing valleys; and yet other times when the whole Lake District was more or less buried under an extensive ice-sheet. Arbitrarily, the landforms attributable to each phase can be described separately.

Corries are armchair-shaped basins in mountain sides, perched high above the valley-floors and surrounded except at their open front by more or less precipitous crags. Examples are the Blea Water and Small Water corries below High Street, or those containing Goat's Water and Low Water below Coniston Old Man. They are the localities where snow first accumulated and became compacted into glacier ice early in a glaciation and they gradually acquired their characteristic shape by glacial modification of some previous gully. Corrie glaciers deepened their beds by abrasion, using rubble frozen into their soles, by bursting off projecting rock fragments, and by plucking away loosened, joint-bounded blocks which became frozen to them. They also enlarged their containing corries in plan by sapping back at the base of the corrie walls which thus retreated in a semi-circular form while maintaining their steepness. Adjacent corries came to impinge on each other leaving knife-edged ridges between: Striding and Swirral Edges on Helvellyn are fine examples. A number of corries contain tarns, including those mentioned at the beginning of this paragraph. Blea Water, which has the remarkable depth of 207 feet, is in a hollow eroded by its glacier out of solid rock, but several tarns, like Scales Tarn below Saddleback, are impounded at least in part by moraine—the rubble shed by the ice at its downstream terminus. The finest of the Lake District's mountain scenery is associated with corries and their tarns and sheer crags, broken by rocky gullies, clefts and chimneys. Most corries open in directions between east and north, presumably because it was south-westerly winds which brought snow and drifted it into hollows on the lee sides of the mountains where it nourished corrie glaciers. The Helvellyn range, like others, is asymmetrical, scalloped by a row of corries on its eastern side but with a relatively less steep descent to the glacial trough of Thirlmere on the west.

For as the climate became colder and the snow-line descended, the corrie glaciers contributed ice to valley glaciers which, also receiving tributaries from plateau ice-fields and from side valleys, flowed out of the mountains into the lowlands. They too eroded their beds, greatly deepening their valleys into steep-sided troughs—the present dales. The glaciers were unable to accommodate themselves to bends in the old valleys and planed off the noses of spurs to make towering crags like Raven Crag, of rhyolite, in Great Langdale, so leaving straighter valleys. The rock of the valley-floors was selectively and irregularly eroded according to the resistance offered. Less indurated rocks, and those weakened by dense jointing or by cleavage, had basins excavated in them, which after the ice melted were occupied by the long ribbon lakes. Wast Water, the deepest, has a maximum depth of 258 feet which takes it 58 feet below sea-level. Windermere consists of two deep basins with shallows between, where islands emerge off Bowness.

If this is a district of crags and lakes it is also one of waterfalls, and the largest are those which pour out of corries and tributary glacial troughs down to the floors of main valleys. The glacier ice of those valleys was more erosive than that of their smaller tributaries, leaving the floors of the latter opening high in the walls of the former. Borrowdale has an array of falls, including Lodore, from such hanging valleys. At temporary termini of these valley glaciers terminal moraines were constructed from debris delivered by ice, and Coniston Water and Windermere have their surface waters dammed behind such moraines at Nibthwaite and Newby Bridge respectively. These moraines belong to a stage in the retreat of the ice, or to a minor readvance, after the maximum.

As the thickness of ice increased, valley glaciers overflowed low cols in the divides and sent off distributary tongues which themselves eroded troughs like the one linking Great and Little Langdale past Blea Tarn. Ice congestion in the northern Lake District caused surface ice from the Thirlmere and Brothers Water glaciers to flow backwards over the cols at Dunmail Raise and Kirkstone Pass respectively, imparting to each of these its glaciated form. Finally the ice covered the high plateau uplands to form a sheet dispersing from an area of maximum accumulation in the Sca Fell area. Erratic stones of tuff and lava were carried outwards on to the Skiddaw Slate up to altitudes of at least 1,800 feet. On the high fells, too, rock was selectively eroded so as to leave hollows, now occupied by small tarns or infilled with peat, interspersed among rocky mounds. These last are perhaps the most characteristic feature of these and other glaciated mountains. They are smoothly moulded by abrasion and are ice-scratched on the up-ice side but are quarried into irregular low crags down-ice because it was on this lee side that the moving ice could shift away loosened, joint-bounded blocks which had become frozen to it. These asymmetrical knolls and hillocks, called *roches moutonnées*, are as abundant on the lower Silurian fells, on valley-sides and on valley and corrie floors, as they are on the plateaux. The islands of the Windermere shallows and of the head of Ullswater are examples just keeping their heads above water.

Remember that it was glacier ice which created lakes, and corrie and plateau tarns; the crags of corries, valley-walls and *roches moutonnées*; waterfalls and cascades. It transported large boulders and distributed them widely. It eroded selectively to produce a rocky, irregular landscape. It enormously deepened the valleys and corries, and in so doing created relief amplitude without which the expression of difference in rock type and so of scenery between the Skiddaw Slate, Borrowdale Volcanic and Silurian areas would have been muted. Glacier ice made the National Park.

Whilst in the mountainous central Lake District ice was mainly erosive, in the lower surrounding country it deposited sheets of till. This is the mixed debris of boulders, sand, silt and clay, eroded and transported by the moving ice and plastered by it over the underlying rock often to a depth of 50 feet or more. It was in many areas moulded by the ice which deposited it into drumlins—oval, stream-lined hills up to half a mile in length and 50 feet high whose long axes accord with the direction of ice-flow. Swarms of drumlins occur in the Crake Valley below Coniston Water, and west of Penrith. Till also occupies parts of the floors and lower slopes of the valleys in the central Lake District, where it underlies much of the farmland enclosed by walls made of stones in part cleared from the surface of the till.

When the ice melted it discharged enormous volumes of melt-water which flowed on, in, under, alongside and away from the disintegrating glaciers, eroding melt-water channels in places where streams would otherwise have no business to be—most small valleys among the fells, if now without a stream, were formed in this way. Melt-water also carried gravel and sand, newly released from the ice or picked up from till beneath it. Some of this material was banked against lingering stagnant ice in the valleys, as was, for example, a gravel terrace 25 feet above the west shore of Windermere; the eastern riser of the terrace is the cast of a wall of the ice then occupying the lake site. Gravel carried by melt-water rivers down-valley away from the decaying ice choked the valley-floors and was built as deltas out into the lakes, producing the characteristic flat alluvial bottoms of the glacial troughs.

The fresh-looking hummocks of rubble, 30 or so feet high, so conspicuous in the corries and valley-heads, are moraines associated with the disintegration of the small glaciers which reoccupied these places in the brief cold spell of just over 10,000 years ago. To this and earlier cold spells, after the main ice-sheet decayed, also belong most of the screes which form aprons below crags. They accumulated when repeated frosts, by causing expansion of freezing water in joints and cleavage planes, broke off fragments from rock faces above.

Since then there has been little change in the terrain. Time has been too short and the erosional and depositional power of modern streams so slight in relation to what went before. It is as if glacier ice left the Lake District only yesterday. Yet there has been change, for, as the climate improved, forest reclothed the valleys and mountains and hid the small features which now contribute so much to the landscape; the *moutonnée* crags, screes, boulders, channels, torrents and moraines are usually smaller in scale than a respectable oak. Man, by clearing the forest, exhumed the glaciated landscape. And it is this intimate scale of the topography of the Lake District which now poses special problems in its conservation.

Geological Maps

On the quarter inch to one mile scale, the sheet covering most of the Lake District (north of an east-west line through the lower ends of Coniston Water and Windermere) is No. 3. The southern area is covered by Nos. 5 and 6 (combined in one sheet).

On the one inch to one mile scale, published sheets exist as follows: 22, Maryport; 23, Cockermouth; 28, Whitehaven; 37, Gosforth.

It follows that although one-inch cover exists for the West Cumberland coalfield and the northern and western fringes of the National Park, for much of the Park it does not.

3
Vegetation and Plants

by R. G. H. BUNCE

THE vegetation of a region is of great importance not only for its intrinsic interest, but also because it provides, directly or indirectly, food and shelter for most other forms of wild life. The natural boundaries of a region are also usually reflected in the vegetation they contain. In the following discussion, after a general introduction, five major zones are described, starting from the seashore, working up to the mountain tops, and ending with the lakes.

The vegetation of the Lake District, as it appears today, is the result of an evolutionary process that has been taking place over thousands of years since the last Ice Age. The region is in the extreme west of Europe and the species present reflect this position, many of them being found only in the Atlantic fringe. The particular parts of the flora which show this feature are the mosses and liverworts in which the Lake District is particularly rich. Other plants accepted as common in Britain, such as Bluebell (*Endymion nonscriptus*) and Foxglove (*Digitalis purpurea*), are oceanic and are rare in a European context.

The combinations in which the species occur are due to interaction between the environment and the activities of man. Whilst the major types of vegetation are determined by environment, the details within each of the major zones have been determined by man's activities, to a greater or lesser extent depending upon the degree of cultivation of the land. Elsewhere in this guide the geology and history of the Park are discussed; the former in particular has a direct effect on vegetation. The relative importance of environmental factors in determining vegetation is difficult to establish; it is for the ecologist to understand these relations and to single out those that are important.

Early man started here by clearing the forests at their upper limits, but by Roman times much of the forest still remained in the valleys —as shown by the position of the roads such as High Street upon the mountain ridges. With the Viking invasions much forest was cleared, as shown by numerous Norse place-names such as those ending in *-thwaite*, meaning a clearing in the forest. During the monastic period and in Elizabethan times the clearing became more rapid, until it was necessary to pass a statute to conserve the woodlands. Forests were exploited mainly for charcoal, which was used to smelt local ore as well as in the mines for pit-props. Until the nineteenth century the woods were intensively used, but since then, the area of woodlands has been building up again.

The development of sheep farming in the nineteenth century was a major change in the ecology of the region, and it is possible that the present decline in agricultural intensity in the uplands may be of a similar magnitude. Concurrent with this decline has been the expansion of coniferous plantations which have altered the appearance of some valleys, although their influence on the wild life has often been overstated.

Historical events such as the Great Depression have had their influence upon the vegetation of the Lake District as it is today. Whilst such trends are important, it is the future that must be examined, for vegetation is dynamic and ecology must be understood as a living and developing science. Conservation measures need to be positive, looking to the future rather than regretting the past; otherwise the idea is encouraged that everything was once better than now and that one should aim to return to some mythical Golden Age, of course unattainable.

Today probably the biggest influence on vegetation is farming, although housing, road building, quarrying and afforestation all have profound effects. Farming practice, through the cultivation of the land and by the control and intensity of grazing animals as well as the use of insecticides, herbicides and fertilisers, has far-reaching effects. Some insecticides, for example those based on chlorinated hydrocarbons, do not break down after application but remain in the soil and can accumulate in the tissues of animals until dangerous levels are reached; beneficial insects are also killed in the initial spraying. Selective herbicides change the balance of species; fertilisers may accumulate in lakes, through run-off of drainage waters into the rivers. There is much concern about pollution, with modern farming becoming increasingly industrialised.

Other major changes may be caused indirectly by external economic factors forcing farmers into practices which they know to harm the environment, in order to maintain the profitability of their farms. For example, changes in sheep and cattle subsidies can alter the balance between arable and pastoral farming systems. The effects of entry into the Common Market, where, in the Mansholt Plan, a policy has been laid down under which land will be progressively taken out of agriculture, may be important, although regional grants may cushion its full impact.

Quarrying, mining and building have a more obvious, catastrophic

Vegetation and Plants

effect on the native vegetation, in that entire habitats are destroyed and only partly replaced with new ones. These, however, are colonised by plants usually different from those that were there originally. Many old quarries contain rare and interesting plants. The effects are therefore not entirely negative, although losses are invariably through vegetation that has been developing for thousands of years being replaced by types that could, if necessary, be created *de novo*. The intrusions of many schemes are mainly upon the appearance of the countryside and should not be confused with their effects upon the vegetation. For example, a line of pylons can destroy a viewpoint but has little or no effect upon what grows. The demand for water from the big cities has led to valleys being flooded and presents a problem that has yet to be solved.

The superficial appearance of the Park is misleading in that the Lake District is famous for its mountains and one might therefore expect its vegetation to consist of montane types. In fact, only a minor proportion of the Park is mountain; almost 65 per cent of the land is below the 1,000-foot contour, emphasising that the greater part of the Park may not be termed mountain and the chief elements of vegetation are not likely to be montane types. In the discussion below, the non-mountain areas are given due weight.

The vegetation of the Park is described under five major headings representing easily recognised zones: Marine; Agricultural (p. 21), Fell (p. 26), Mountain top (p. 32) and Aquatic (p. 32). Some plants may overlap all five zones but each group contains characteristic species and vegetation types. These types are based on observation rather than survey and form a framework around which the vegetation of the Park may be described.

Each section is divided further into different types which may be identified by species selected as indicative of the particular habitat in which that type is found. In the table given at the end of the chapter (p. 34), pairs of plants particularly characteristic of the main types described in the text are listed.

Marine zone. The habitats on the seashore are included in this section, as well as others significantly affected by salt water. Only a small area of the National Park extends to the shore, but the maritime vegetation represents an interesting contrast with that found in the uplands. Although a very small part of Morecambe Bay is included in the Park, the main section runs from Gutterby Spa to just north of the Ravenglass estuary. The beaches have gently sloping surfaces, patches of sand alternating with pebble and boulder beds. The large tidal range and shallow beaches bring about wide ranges for different shore communities. In the deep water, beds of Kelp (*Laminaria* spp.) are probably present, as fronds are often found among the debris washed up on the beach. Higher up the beach, growing on the pebble and boulder beds, the range of seaweed species is rather limited, probably because of the turbidity of the water, but Bladder Wrack (*Fucus vesiculosus*) and Toothed Wrack (*F. serratus*) are widespread. Rock pools are relatively few and not particularly rich in species.

At the limit of the highest spring tides the first flowering plants are found, often growing amongst the shore debris, where Curled Dock (*Rumex crispus*) is amongst the first species to be found. In the Gutterby area the beaches are backed by low cliffs, consisting of glacial drift comprising compacted mud, stones and boulders. These cliffs have a characteristic vegetation, often containing groups of Gorse (*Ulex europaeus*) and other flowering plants such as Sea Campion (*Silene maritima*) and Red Fescue (*Festuca rubra*). Frequent gullies have been cut back into these relatively soft cliffs and much slumping is caused by erosion by the sea, giving rise to highly unstable conditions and much bare soil.

In the Ravenglass estuary two different types of coastal vegetation occur. Firstly there is a small area of saltmarsh, where the substratum consists of fine mud and there is a gently sloping shoreline. The small section of Morecambe Bay that comes in the Park is also of this type. Although limited in extent, many of the typical saltmarsh species are found, such as Sea Arrow-grass (*Triglochin maritima*) and Sea Spurrey (*Spergularia maritima*). On the seaward side of the marsh the mud is bare and is colonised at first by isolated plants of Glasswort (*Salicornia* spp.); with the accumulation of mud other plants become established. The middle area of the marsh is dissected by many channels with bare mud, but eventually in the upper part the vegetation forms a complete cover.

Secondly there are the sand-dunes at Ravenglass and Eskmeals: well developed and excellent examples of this type of vegetation. The yellow dunes nearest the sea still have highly mobile sand, with Marram Grass (*Ammophila arenaria*) and Lyme Grass (*Elymus arenarius*) colonising the bare surface and helping to stabilise the dunes. Behind the foredunes are depressions, called slacks, which are often filled with water in winter and have an exceptionally rich flora, containing a number of rarities. Plants such as Creeping Willow (*Salix repens*) and Marsh Pennywort (*Hydrocotyle vulgaris*) are indicative of this habitat. Rarities present include two species of Adder's Tongue (*Ophioglossum* spp.) and Dune Centaury (*Centaurium littorale*).

Agricultural zone. This zone for present purposes is defined as the area capable of being ploughed, *i.e.* it is arable land, and is situated below the open fellside, often delineated by the mountain wall. Contained in this category is a wide range of land, from the coastal plain to sheltered valleys and to the marginal hill-land usually below the 800-foot contour. Within the area some land, because of rock being on or near the surface, cannot be ploughed, but much of the remainder can carry crops, such as barley or potatoes. Further, man has concentrated his activities in land improvement within this zone to increase crop yields as well as living mainly within it. Not only has the landscape therefore been highly modified, but also the flora much altered. Nonetheless this zone forms an appreciable proportion of the Park and provides the contrasting background against which the more severe hills may be appreciated.

In low-lying, poorly-drained areas marshland is present, although as a habitat it is always under pressure from land improvement projects and was once far more extensive than may be seen today. For example, the old coach road from Kendal to Ulverston did not take the present route across the flat land on the edge of Morecambe Bay, but followed an inland course to avoid the marshes. In the coastal plain in the west, and in more limited areas inland, for example near Rusland, various willows, particularly Common Sallow (*Salix atrocinerea*), have spread on to the marshes and formed scrub that is particularly rich in bird life. The flora of such areas is often diverse, with Flag Iris (*Iris pseudacorus*) and Bottle Sedge (*Carex rostrata*) growing in the water. At North Fen, a Nature Reserve to the north of Esthwaite Water, various stages in the colonisation of open water may be seen. Elsewhere this type of vegetation is restricted by the limitation of the habitat through drainage. Small groups of Alder trees (*Alnus glutinosa*) may be seen around ponds, on the margins of which plants such as Brooklime (*Veronica beccabunga*) and Water Starworts (*Callitriche* spp.) may be found. These small patches of native vegetation provide cover for many insects and small birds, contributing much to the diversity of lowland farm country.

Major features of the lowland landscape seen by motorist and walker alike are the hedgerows and verges. In intensively farmed areas these form an invaluable refuge for all forms of wild life. Whilst very common in lowland Britain they are of restricted distribution in Europe.

The composition of the hedgerows varies according to underlying geology, drainage, management, and age of the hedge, and can often give clues to the likely plants that would be present if the influence of man were removed. On alluvial soil in the lowlands, the hedges are likely to consist of species such as Hazel (*Corylus avellana*) and Ash (*Fraxinus excelsior*) with Bramble (*Rubus fruticosus* agg.) and Wild Rose (*Rosa* spp.) scrambling about amongst the bushes. On the shallower, stonier and nutrient-poor soils occurring on the slates and volcanic rocks there are fewer species and Hawthorn (*Crataegus monogyna*) is more widespread. Bird Cherry (*Prunus padus*) occurs in both upland and lowland hedges and is a feature of the north, being absent in the south of England. The poorer hedgerows also have different species growing beneath them, Foxglove and Wood Sage (*Teucrium scorodonia*) as opposed to Goose Grass (*Galium aparine*) and Red Campion (*Silene dioica*) in the lowlands. The hedgerows that occur on the Carboniferous Limestone are different again, being particularly rich in attractive species such as Wood Geranium (*Geranium sylvaticum*) and Field Scabious (*Knautia arvensis*).

The road margins, including both verges and hedge-banks, also vary with the underlying geology but have more obvious differences due to management and their age—new verges being seeded with uniform mixtures of only one or two species. Some produce superb displays of colour. The practice of spraying road verges, which has greatly affected the composition of species in some counties, is fortunately rare in the Park. Also many roadsides are still maintained by local roadmen who take a pride in their work, adding greatly to the attractiveness as well as to the richness of their flora. There is 'a succession of flowers starting in spring with Celandine (*Ranunculus ficaria*) and Dandelion (*Taraxacum* spp.) and ending in late summer with Hardhead (*Centaurea nigra*) and Yarrow (*Achillea millefolium*). On the more lowland verges, many plants are similar to those found in the meadows, described below, whereas those on the acidic soils are often grazed, and have mountain grassland species. Limestone verges are again different, having a rich flora similar to that of grassland nearby. Often the species seeded into new verges are unsuited to the soil and are gradually replaced by local plants.

To many observers, one of the most attractive features of the Lake District countryside are the woodlands. Most of these are in the agricultural zone, but although extending on to the fells they are such a well-defined habitat that they are discussed together. The woodlands of the region have a long history of intensive use; there is probably more woodland today than for the last seven hundred years, as well as less industrial activity within it. Before the fifteenth century the forests were cleared for grazing and increasingly the wood was used for charcoal production as well as for many minor

PLATE I. The view of Derwent Water from Latrigg is as popular today as with the early tourists of the eighteenth century

PLATE II. The craggy scenery of the Coniston Fells, Langdale, and Borrowdale, and distantly, Derwent Water and Skiddaw

PLATE III. Lake-shore trees, forest, and bare fellside; looking across Coniston Water at the Old M

PLATE IV. Red deer grazing in Grizedale Forest

industries such as bobbin-making and tanning. The woods were cut on a cycle of about seventeen years and consequently there were relatively few mature trees. In the early nineteenth century the techniques of forestry, as understood today, began to be practised in the Lake District; new woods were planted with oak at first, but later conifers such as Larch (*Larix* spp.) and Scots Pine (*Pinus sylvestris*) were introduced, culminating in the large plantations made since the First World War by the Forestry Commission mainly of Sitka Spruce (*Picea sichensis*) in valleys such as Ennerdale and Grizedale.

Firstly, therefore, the woods comprising mainly non-indigenous species may be considered separately. These are often ignored by naturalists as containing little of interest. Nevertheless they contribute to both flora and wild life of the Park as well as to its appearance. Most of the forests have been planted on the lower fells, but many are also within the agricultural zone, although on the poorer soils. Many of the plantations are still young with few plants in the dense shade and have not had time to develop a distinctive flora. In the rides the original vegetation remains much as before, although altering in relative quantities of species, often through the removal of grazing. Some of the older plantations, especially larch, have shade-bearing plants, such as ferns and mosses, growing within them. The cuttings and embankments of the road systems have also a developing vegetation cover which may in time become more interesting.

The real interest of the naturalist is however in the native woodlands, which may be divided into four types occupying different habitats, although they frequently overlap. The first type, now of rather limited extent, occurs on very wet soils, either on valley-bottoms or by stream-sides. The tree cover is mainly Alder and Ash with many species that require wet soils, such as Meadow Sweet (*Filipendula ulmaria*) and Water Celery (*Oenanthe crocata*). Vegetation of this type may be found by the river Duddon and by many small streams elsewhere. Among the mountains, for example in Martindale, there is an upland type, growing between about 1,000 and 1,400 feet on wet clay soils. Alder again forms the canopy, but many species from the surrounding grassland are present and the woods are heavily grazed, whereas the lowland woods usually are not.

A more widespread type in the valleys is the lowland, mixed deciduous woodland, with Pedunculate Oak (*Quercus robur*) and Ash as the main canopy species. Hazel is frequent in the understorey and there is a well-developed ground flora, usually dominated by brambles but with many spring-flowering species such as Bluebell and Wood Anemone (*Anemone nemorosa*). The soils are usually deep and moist with a relatively high nutrient status. Wild Daffodil (*Narcissus pseudonarcissus*) grows in this type of wood, although most daffodils in the Lake District are naturalised cultivated varieties.

In contrast the most widespread woodland type in the Lake District occurs on shallow, often well-drained soils originating from the slates and volcanic rocks that are usually acidic and low in calcium. Here the main tree species are Sessile Oak (*Quercus petraea*) and Birch (*Betula pubescens* and *B. verrucosa* and hybrids). There are many variations depending upon the type of bedrock, angle of slope, soil depth, and rainfall. In the Borrowdale Valley, and to a lesser extent in Buttermere and Ennerdale, the high rainfall results in high humidity and has led to the development of an extremely varied moss and liverwort flora, indeed amongst the richest in western Europe. Many of these woodlands are grazed and used for shelter by sheep and cattle. The ground flora is characterised by an abundance of mosses and plants such as Wavy Hair Grass (*Deschampsia flexuosa*) and Bilberry (*Vaccinium myrtillus*). Woods of this type extend to about 1,400 feet in the Newlands Valley in the Keskadale and Birkrigg oakwoods, but generally are found at lower elevations. Around the lakes ornamental conifers have often been planted in such woods, but beyond altering the appearance they have little effect.

The last main type is that limited to limestone areas. Here the soils are usually shallow but are high in calcium and support a rich flora which contains many rare species. Ash is the most widespread tree, although Yew (*Taxus baccata*) is often found on rock outcrops. Hazel in many places forms scrub, in the open, with other shrubs such as Spindle (*Euonymus europaeus*). Beneath dense trees the ground flora is often sparse, with Dog's Mercury (*Mercurialis perennis*) and Wood Brome (*Brachypodium sylvaticum*) characteristic. In glades and on woodland margins, however, the ground vegetation is usually luxuriant and rich in species—many of the rarer species grow only in semi-shade, for example Red Helleborine (*Epipactis atrorubens*) and Fingered Sedge (*Carex digitata*).

Concern is often expressed at the virtual lack of regeneration in Lakeland woods. But many are relatively young in the lifespan of trees—mostly eighty years old, or younger, whereas the oak has a potential of several hundred years. Dead trees are uncommon, usually smaller specimens. In consequence, many of the woods have relatively dense canopies and so are not suitable for regeneration. The native woodlands are certainly not moribund, for in many areas in Lakeland today the trees are recolonising land, upon the reduction in grazing pressure following the decline in farming intensity. In the Coniston and Seathwaite Valleys woodland is present in places now that were shown to be bare of trees on maps only 50 years old.

The woodlands are often on land unsuitable for agriculture, and together with marshy areas form the bulk of the uncultivated land in the agricultural zone. The remainder of the land is mostly pasture, either permanent or temporary, and has been ploughed at some time in the past; the high rainfall favouring a pastoral pattern of agriculture. On the ploughed land, of which many farms have one or two fields mainly to produce animal foodstuffs, potatoes are sometimes grown as cash crops. Although containing few unusual plants, these fields have a characteristic flora of weed species, such as Chickweed (*Stellaria media*) and Couch Grass (*Agropyron repens*). They add to the diversity of appearance in an otherwise uniform pastoral area and form a source of food for birds and animals.

The arable crops are often alternated with grass. These reseeded leys, important in modern grassland husbandry, contain species such as Rye Grass (*Lolium perenne*) and Cocksfoot (*Dactylis glomerata*) that have been improved in recent years by breeding programmes. The swards are highly productive and often heavily fertilised. Few species usually are included in the original mixture and such grasslands are not diverse, in contrast to the traditional permanent pasture with its close-cropped turf and many non-grass species, the proportions varying according to grazing regime and soil type. Although below full cropping potential in agricultural terms, grasslands of this type are amongst the most diverse habitats to be found, in terms of animals and plants. If they are grazed badly, e.g. at the wrong time of year, too lightly, or too intensively, then the balance is upset and coarse grasses and weeds such as docks and thistles become widespread. Similarly, if not cut regularly, Bracken (*Pteridium aquilinum*) will often spread rapidly into such fields, as will rushes if the drains are neglected. Pastures on the rocks with acidic soils require liming to maintain their fertility and contain much Bent Grass (*Agrostis tenuis*) and Yorkshire Fog (*Holcus lanatus*), with other species such as Self-heal (*Prunella vulgaris*) and Daisy (*Bellis perennis*) often present. On the limestone pastures, such as those on Whitbarrow, the flora is very different, rich in colourful flowering species like Rockrose (*Helianthemum chamaecistus*) and Salad Burnet (*Poterium sanguisorba*). The main grass species is Blue Grass (*Sesleria decumbens*). The area is botanically interesting with many rarities such as Hoary Rockrose (*H. canum*), and insects are plentiful.

Although not strictly a grassland habitat, no discussion of the limestone areas would be complete without mention of the limestone pavements, found for example on Whitbarrow. These are areas of rock bared of soil by glacial action and subsequently colonised only by vegetation in the cracks, known as grikes, where the water has eroded the rock. The grikes have within them many species usually associated with woodland, such as Hart's Tongue Fern (*Phyllitis scolopendrium*) and Dog's Mercury, being present as well as rarities which grow only in grikes, such as Rigid Buckler-fern (*Dyropteris villarsii*). The walls are also a feature of the landscape, often overgrown with Ivy (*Hedera helix*) and with many characteristic mosses.

A type of vegetation closely related to the pasture, fast disappearing, is the traditional hay meadow. These grow on the soil types most amenable to reseeding and agricultural improvement. The old hay meadows contain tall-growing grasses and other plants such as Ox-eye Daisy (*Chrysanthemum leucanthemum*) and Sorrel (*Rumex acetosa*) that immediately before hay-time present an attractive display of flowers. Heavy application of nitrogenous fertiliser can have a similar effect to reseeding, as the grasses respond vigorously to the increased nutrient levels. In the higher valleys, often where there is some downwash of enriched water, are small patches of sub-alpine meadow containing species found in similar habitats in the Alps, besides species with superb flowers such as Globe Flower (*Trollius europaeus*) and Melancholy Thistle (*Cirsium heterophyllum*). Similar vegetation may be found on comparable soils on mountain cliffs.

Fell zone. Above the agricultural zone, most of the land is open fell and is the landscape that most visitors have come to see whether on foot or by road. To the lowlander it is the hills with rocky cliffs, green slopes and screes that differ most from his familiar rolling landscapes. One of the most striking features of the lower fells is the walls, which mark farm boundaries and form the units of management used by the farmers. These walls have their own distinctive flora, with mosses and lichens most common, but occasionally other plants such as Common Polypody (*Polypodium vulgare*) and Parsley Fern (*Cryptogramma crispa*). Many of the higher walls are in ruins but the lower ones are still generally well looked after.

The superficial appearance of the vegetation of the fells is dominated by sheep, although the main types of vegetation present are determined more by the nature of the soil than by grazing. The soil is in turn the result largely of interaction between the material from which it has been formed and the patterns of local drainage.

Vegetation and Plants

The sheep, in such numbers as at present, are relative newcomers to the hills, taking over from cattle as the commonest grazing animals in the nineteenth century. It is difficult to pinpoint vegetational changes. Sheep are highly selective grazing animals, cropping the herbage closely, ignoring the coarser grasses; cattle are more catholic in their tastes and help to control coarser species by trampling. The increase in sheep numbers in some areas for economic reasons is causing concern that some upland pastures are being overgrazed. Vegetation on ungrazed rock ledges is much taller and contains species not found in the grasslands below; this difference, although in part due to habitat, indicates the effects of grazing on vegetation. In general heavy grazing favours grasses, whereas under lower intensity more woody species are able to survive, even though they may be only dwarf shrubs such as heathers.

Much of the vegetation on the fells is less obviously disturbed by man than in the lowlands, in that ploughing and fertilising have affected only a small area. The major influence is indirect—through the use of grazing animals. Whereas some mountain cliff vegetation is virtually unaffected, except by rock climbers, the remainder of the hillsides are far removed from their undisturbed state. To this extent the smooth green hills are as much a product of man as the lowland patchwork of fields. Recent advances in the techniques of reclamation can greatly alter the constituents of upland grasslands but have often been agriculturally successful only in the short term, although altering the balance of species for many years. Once disturbed the balance can rarely be restored, and in the long term such reclamation may often be deleterious.

The major type of vegetation on the lower fells is rough grassland, much of which may be invaded by bracken. These slopes are generally well drained and although often poor in nutrients have only a shallow layer of undecayed humus on the surface, beneath which is mineral soil. The commonest grasses are Sheep's Fescue (*Festuca ovina*) and Bent Grass—which provide the best grazing on the mountain. Other plants such as Heath Bedstraw (*Galium saxatile*) grow amongst the grasses and there is often a carpet of mosses. In areas where the bracken is most vigorous there is no ground cover at all as the dead fronds fall thickly to the ground. In more open areas, however, many of the grassland species are present and occasionally woodland plants, such as Bluebell, are able to continue growing under the fronds after the woodland cover has long been removed. Although bracken-covered hillsides are extensive, they are often broken by small rocky outcrops and wet patches, the vegetation of which is discussed below. Bracken used to be controlled by crushing the young fronds with an iron bar towed behind horses, as well as being cut by scythe for bedding. That is outdated, and cutting by tractor as now practised is less effective; in recent years, therefore, the bracken has spread, and seems likely to continue to do so, although trials with a new herbicide are in progress.

Where the soils are even poorer in nutrients but remain well drained, heath-type vegetation predominates. Such heaths are found frequently on the Skiddaw Slates in the north of the Park and also widely on the eastern fells. Ling Heather (*Calluna vulgaris*) and Bilberry are the main species and the relation with the previous type is complex, depending to a certain extent on grazing pressure. These moors are often associated with grouse, as the birds feed largely on heather, and in consequence the moors are burnt in rotation to provide shoots of suitable ages. The heather moors are poor in plant species, but the purple colour of the hills in late summer and early autumn is very striking and indicates the affinities of the Lake District fells with the heaths of south-west England, North Wales, and Scotland. As the drainage becomes more restricted, so conditions lead to the build-up of peat due to bad aeration stopping the bacteria breaking down the dead plant remains. On the shallow slopes, Mat Grass (*Nardus stricta*) intergrades with the heather, although it can also be favoured by selective grazing, with other species such as Tormentil (*Potentilla erecta*), and replaces it as the main species over large areas. These areas have cold, wet soils with a layer of amorphous peat on the surface, and are poor in both species and productivity. Mat Grass moors are particularly widespread on western fells such as Black Combe, where the hills are more rounded than in the central peaks. On similar rolling hills in the east similar vegetation occurs; it is one of the most widespread types in the Park. In some areas Western Gorse (*Ulex gallii*) and Juniper (*Juniperus communis*) have colonised the grassland, but in general it shows a monotonous type of vegetation.

In wetter but non-stagnant areas, bogs with Purple Moor Grass (*Molinia caerulea*) are found, with plants such as Cross-leaved Heath (*Erica tetralix*) and Bog Asphodel (*Narthecium ossifragum*) present. On hills in the west, as above Miterdale, and on the more rolling hills in the north-east, such vegetation is locally abundant, although it is not so common on the steeper hills in the centre of the Park. The soils are always wet and usually have a thick layer of fibrous peat. As the slopes get shallower, other species take over and bogs with stagnant water, dominated by Cotton Grass (*Eriophorum vaginatum*), occur. This vegetation is characteristic of large areas in the Pennines and has stagnant water present in deep peat that is extremely poor in nutrients. It has a few characteristic species such as Ling Heather and Cloud Berry (*Rubus chamaemorus*) and is most common in the east of the Lake District, occurring elsewhere only in small patches where the drainage is bad. The other species of Cotton Grass (*Eriophorum angustifolium*) overlaps the last two types, and is particularly common on areas of bare peat and on the margins of shallow pools. In both types various species of *Sphagnum*, the bog-mosses, are widespread and important in the build-up of peat. (See also Figure 4.)

Bog vegetation is found not only on the fells; similar conditions may occur in the valleys near Rusland and near Broughton-in-Furness, although they are caused by different processes. There the vegetation originally colonised open water relatively rich in nutrients and led to the build-up of peat. As the peat increased so the supply of nutrients from the ground diminished until eventually the top layer of the peat became acidic. In these bogs (or rather mosses, as they are called) the vegetation therefore is similar to that on upland areas with conditions favouring peat formation, although when drying out Ling Heather becomes abundant, and the bog surface is colonised by birch and pine.

In autumn and winter these broad types of vegetation may be picked out by the different colours they exhibit on the mountains. The rusty red of bracken-covered slopes is quite distinctive, in contrast to the olive green of the grasslands with Sheep's Fescue and Bent Grass. Rush areas show up usually as strips along seepage zones, whereas Mat Grass moors show up white. The red-brown flatter areas contain bog vegetation. Trees may also be picked out by colour differences, oak being dull brown; birch, reddish-purple; ash, grey.

In contrast to the above types, which are usually associated with relatively deep soils, the vegetation of rocky habitats grows on shallow, skeletal soils or even in rocky cracks where there is no soil at all. The vegetation is in patches separated by areas of bare rock, although that is often covered by lichens. Most striking of these habitats are the cliffs, the vegetation of which varies according to their altitude and geology. At low levels in the valleys, outcrops such as Shepherd's Crag in Borrowdale often have quite large trees growing on them and are highly vegetated. Cliffs consisting of mainly acidic rocks, such as the Borrowdale Volcanic Group, have different vegetation from the limestone crags in the south of the Park. The former have Oak and Rowan (*Sorbus aucuparia*), often growing directly out of cracks in the rock, whereas the latter have Yew and a variety of limestone trees and shrubs such as Privet (*Ligustrum*

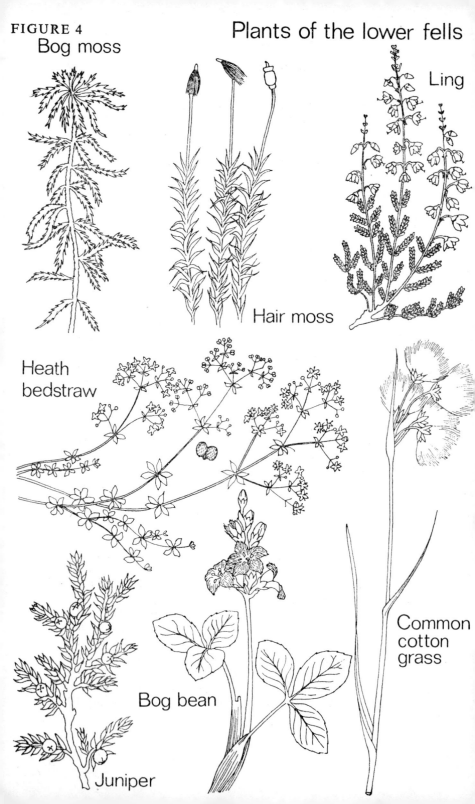

FIGURE 4
Plants of the lower fells

vulgare) and White Beam (*Sorbus aria*). On the acidic cliffs the ledges have Bell Heather (*Erica cinerea*) and Honeysuckle (*Lonicera periclymenum*), with Ivy often scrambling up them. Limestone cliffs also often have Ivy in addition to other distinctive plants such as Horseshoe Vetch (*Hippocrepis comosa*).

Trees are able to survive to much higher altitudes on cliffs than elsewhere, for example up to 2,000 feet on St. Sunday Crag in the Helvellyn range, indicating that the present tree-line is not merely climatic but also related to historical factors. The vegetation on the higher cliffs contains interesting combinations of species from a variety of habitats, from the dry acid soils on the buttresses to the almost aquatic conditions in the wet gullies. Many plants found on the crags are not found in the surrounding grasslands. At quite low elevations occur some of the most distinctive of the Lake District plants—the arctic-alpines. These are plants which grow in the Arctic at sea-level but farther south are found only on mountains, as they are not able to compete in the dense vegetation of the lowlands. They occur scattered through various types of vegetation and have a particular interest due to their distribution as well as to their usually small size combined with highly attractive flowers. On the lower cliffs Alpine Lady's Mantle (*Alchemilla alpina*) and Roseroot (*Sedum rosea*) may be found, but these are more frequent on the higher cliffs with rarer species such as Mountain Sorrel (*Oxyria digyna*) and Purple Saxifrage (*Saxifraga oppositifolia*). Other interesting plants present include those more usually associated with meadows such as Meadow Sweet and Globe Flower. The actual vegetation types in which these species occur depend largely upon the nutrient status of the soil and the amount of water present. On the dry buttresses are many species common to the heaths described above, and these intergrade into ledges more similar to the meadows; and finally the water trickles have vegetation related to the mountain springs described below.

Often beneath the cliffs there are areas of scree, loose mobile rock that has fallen from above as a result of erosion. These screes are freely drained and so, although the rainfall is high, are very dry. One plant in particular, Parsley Fern, is adapted to growing on screes; other species, such as Bilberry, are also able to survive, and in some areas a succession of species may follow, colonising the original bare scree. However, the large screes at Wast Water and Ennerdale, for example, are colonised very slowly: the rate depending upon the size of the blocks and the mobility of the scree. Large areas are still covered only with lichens, with mosses of the genus *Rhacomitrium* growing mainly in the hollows between boulders. In some areas the screes have been completely covered by vegetation, although the shapes of the blocks can still be seen. Patterns can also be seen in the colonisation of scree, usually taking the form of tongues of vegetation extending downwards.

Some erosion, due to badly placed and constructed footpaths, can lead to new scree being formed. At present these areas are limited, but there is likely to be a problem in the future with increasing indiscriminate use of some footpaths. Sheep tracks too can cause erosion, as can scree running. Climbers also destroy cliff vegetation by 'cleaning up' the rock, and the continuing expansion of the sport presents a potential threat.

Mountain top zone. The actual mountain tops, defined as the area where the summits flatten out, occupy under 5 per cent of the Park, and represent the most extreme climate in which plants have to grow. The low temperatures experienced on the summits, combined with exposure to wind, mean that the growing season is very short and is comparable to a sub-arctic type. The vegetation shows many features in common with the Arctic in that it is discontinuous, grows slowly, and consists mainly of long-lived perennials. Exclusion of grazing has relatively little effect at these altitudes, the growth being anyway so slow.

On the areas where the rock is near the surface, sub-alpine heath is found with some of the species from the heath described above, and with others such as Cowberry (*Empetrum nigrum*) and Alpine Clubmoss (*Lycopodium alpinum*). Such vegetation is also found on the summits of Scottish hills as well as in Scandinavia and the Arctic. Wavy Hair Moss (*Rhacomitrium lanuginosum*) is an important constituent depending upon the stoniness of the soil and the degree of exposure. Where the ground surface is covered with rock fragments this moss may be the only species present. True Arctic plants such as Dwarf Willow (*Salix herbacea*) and Stiff Sedge (*Carex bigelowii*) can be seen on the hills north of Ennerdale and on Wetherlam in the Coniston range.

In areas where the drainage is poor and in consequence peat has formed, there is vegetation similar to the bog types described above, except that it is dwarfer and even poorer in species. Changes in these areas proceed very slowly.

Aquatic zone. A major feature of an area with such high rainfall as the Lake District is the abundance of small springs that give rise to rivers which in turn feed lakes, all of which are considered in the present section. The mountain springs have a distinct flora that

Vegetation and Plants

differs according to the rocks the water has been passing through on its way to the surface. Most springs have some enriching effects on the surrounding vegetation, but in limestone areas these are pronounced. In the mountain springs various rushes, notably Soft Rush (*Juncus effusus*) and Articulate Rush (*Juncus articulatus*), are often present with a distinctive flora of mosses and liverworts, often with Starry Saxifrage (*Saxifraga stellaris*), an arctic-alpine, growing amongst them. In the calcareous springs the flora is different, highly attractive species such as Bird's-Eye Primrose (*Primula farinosa*) and Yellow Saxifrage (*Saxifraga aizoides*) being present, as well as many other unusual and interesting plants.

The small streams derived from the springs usually have rocky or stony bottoms and plants such as Lesser Spearwort (*Ranunculus flammula*) and Golden Saxifrage (*Chrysosplenium oppositifolium*) associated with them, as well as many mosses and liverworts on the rocks beside them. Often these streams flow into mountain tarns which, together with other isolated pools, are particularly rich in microscopic algae, especially Desmids and Diatoms. In addition, certain vascular plants such as Water Lobelia (*Lobelia dortmanna*) and Small Bur-reed (*Sparganium minimum*) occur.

The streams often pass into ravines or gills, with steep rocky sides, which, as a result of the enclosed space and water, have highly humid atmospheres. The sides of the gills often have enriched water running down to add diversity. The vegetation of these ravines, such as Piers Ghyll on Sca Fell and Dungeon Ghyll in Langdale, resembles that of the mountain cliffs, containing many of the same species but with fewer arctic-alpines and with many more mosses, liverworts and ferns. Many of the bryophytes are western species, Filmy Ferns (*Hymenophyllum* spp.) being particularly characteristic. Trees are often present in a surprising range of species, as in Tilberthwaite Ghyll near Coniston. Such ravines may provide centres from which trees would spread out on to the hillsides again if the grazing pressure were to decline, as may already be seen on the west coast of Scotland.

As the streams run farther down the hillsides, so the gradients decrease and different types of margins are found with different associated plants. The nutrient content of the water also rises as the rivers run through agricultural land and receive enriched run-off from the fields. Here the banks support a rich flora, often containing many species from the surrounding meadows which grow extremely well, indicating their potential when not cut and grazed, *e.g.* Meadow Sweet and Water Celery. The streams are usually lined with Alder, but Ash and Wych Elm (*Ulmus glabra*) are quite common. The gradient declines further until the flood-plain is reached where rivers meander and have muddy as well as sluggish bottoms. Here Common Reed (*Phragmites communis*) and Reed Canary Grass (*Phalaris arundinacea*) grow beside and in the water, extending into the tidal reaches. Other aquatic species such as Water Milfoil (*Myriophyllum* spp.) are in the water. Running into such streams are drainage ditches, in which the vegetation is luxuriant with species characteristic of rich habitats such as Woody Nightshade (*Solanum dulcamara*) and Nettles (*Urtica dioica*).

Finally, there are the lakes themselves. These are of two broad types: firstly those that have a high nutrient content and water containing many suspended silt particles, and secondly those that are poor in nutrients and have very clear water. Esthwaite Water is the prime example of the first type and Wast Water of the second. They have very different floras, both in terms of the algae that live in the water and in the flowering plants that grow on the margins. In the Esthwaite type the margins have a wide range of species such as Reedmace (*Typha angustifolia*) and Common Bulrush (*Scirpus lacustris*), whereas the second type are relatively poor in species, although these are distinctive like Quillwort (*Isoetes lacustris*) and Awlwort (*Subularia aquatica*).

ENGLISH NAMES OF PAIRS OF PLANTS INDICATIVE OF THE MAIN HABITATS MENTIONED IN THE TEXT

MARINE ZONE	Seashores		Bladder Wrack, Toothed Wrack
	Cliffs		Sea Campion, Red Fescue
	Saltmarshes		Sea Arrow-grass, Sea Spurrey
	Dunes		Marram Grass, Creeping Willow

AGRICULTURAL ZONE	Marshland		Common Sallow, Flag Iris
	Hedgerows	(Upland)	Hawthorn, Foxglove
		(Lowland)	Hazel, Red Campion
		(Limestone)	Wood Geranium, Field Scabious
	Verges		Dandelion, Hardhead
	Woodlands	(Plantations)	Larch, Sitka Spruce
		(Riverside)	Alder, Meadow Sweet
		(Lowland)	Pedunculate Oak, Wood Anemone
		(Acidic)	Sessile Oak, Bilberry
		(Limestone)	Spindle, Dog's Mercury

Vegetation and Plants

	Arable		Chickweed, Couch Grass
	Grasslands	(Acidic)	Yorkshire Fog, Self-heal
		(Limestone)	Blue Grass, Salad Burnet
	Meadows		Ox-eye Daisy, Melancholy Thistle
FELL ZONE	Grasslands		Sheep's Fescue, Bent Grass
	Heaths		Ling Heather, Bilberry
	Moors		Mat Grass, Tormentil
	Bogs		Purple Moor Grass, Bog Asphodel
	Blanket bogs		Cotton Grass, Ling Heather
	Cliffs	(Limestone)	Yew, White Beam
		(Acidic)	Rowan, Bell Heather
		(Mountain)	Roseroot, Mountain Sorrel
	Screes		Parsley Fern, Bilberry
MOUNTAIN TOP ZONE	Rocks		Alpine Clubmoss, Dwarf Willow
AQUATIC ZONE	Springs	(Mountain)	Soft Rush, Starry Saxifrage
		(Limestone)	Bird's-Eye Primrose, Yellow Saxifrage
	Tarns		Water Lobelia, Small Bur-reed
	Rivers	(Upland)	Golden Saxifrage, Soft Rush
		(Central)	Water Celery, Alder
		(Lowland)	Common Reed, Water Milfoil
	Lakes	(Clear)	Quillwort, Awlwort
		(Turbid)	Reedmace, Common Bulrush

4
Animals of the National Park

by R. J. ELLIOTT

SOME eight thousand years ago the Lake District was largely clothed in forest. This extended from sea-level to about 2,000 feet O.D., above which forest gave way to montane grasslands and heaths through a transitional zone of shrub. Wetlands, *i.e.* open water, marshland, bogs, fens, carrs, were undoubtedly more extensive than they are today, whilst other habitats such as sand-dunes were probably less so.

Over the centuries this vast expanse of forest has been progressively cleared by man until today only vestiges remain. Likewise the wetlands have been drained and burned, the result being their desiccation and disappearance. The montane heaths and grasslands, and the heaths and grasslands derived from cleared forests, have been modified by the grazing of generations of domestic animals. Many wild animals associated with the original habitats have been exterminated: bear, beaver, elk, bison, lynx and wolf, to name a few.

In lowland Britain the process of human interference has continued beyond the stage of merely modifying the original natural habitats; for the most part these have been wholly destroyed by cultivation and conversion into farmland, town and factory. However, in the Lake District the rugged terrain, predominantly poor soils and the inclement climate of the fells have confined arable farming and human settlement mainly to the valley-bottoms and lower, gentler slopes. Exploitation of the remainder of the area has been largely restricted to its use as rough pasturage for domestic animals. Thus, although burning, draining, and the grazing of stock have all contributed to a gross modification of the original habitats, most of these have survived as essentially natural ecosystems (*i.e.* in the sense that their character is determined more by nature than by man). The conservation of these habitats is of critical importance in safeguarding the unique fauna (and the scenic quality) of the Lake District, for whilst some animals appear to be equally at home in a man-made environment such as arable farmland, others are wholly dependent on natural habitats for their survival.

Animals of the National Park

The mountains and fells. The montane area, *i.e.* land above 2,000 feet, is not particularly rich in the larger animal species. This is most apparent in winter when, apart from possibly a wandering Buzzard or Raven, the tops seem to be destitute of animal life. Indeed in severe weather the only inhabitants to be seen may be an occasional small troop of Snow Buntings. In summer, although larger numbers of animals are evident these are mainly invertebrates, especially insects.

Few species are confined to the mountain tops; those that are, are all invertebrates. The remainder also occur in sub-montane habitats where they are usually more abundant. The Meadow Pipit is a good example. It is the commonest bird of the mountain tops and also of the fells below. It will be found wherever there is coarse vegetation in which to nest and to find the insects with which to feed its young. The only other common breeding bird of the highest ground, the Wheatear, may be encountered on the barest slopes nesting among stones in scree or drystone walls. The Skylark, singing far above the highest peak, although much less common, is perhaps more likely to be noticed by visitors. It nests in rough grassland, like the meadow pipit, up to 3,000 feet.

The richness of animal life, in terms of the number of both individuals and species, tends to increase progressively with decreasing altitude. The Ring Ousel is possibly the most characteristic bird of the sub-montane fells. It takes over from the 'lowland' Blackbird at 750 or 1,000 feet and may be found nesting among scree or in rough vegetation, usually in the vicinity of streams, almost up to the montane zone. But the most spectacular animal of the fells is undoubtedly the Red Deer, the largest surviving member of our native land mammals. A herd of about 300 individuals occupies Martindale, which incidentally represents the only example of 'deer forest' in Britain south of the Scottish Border. These are believed to be direct descendants of animals that roamed the pristine forests of the Lake District and subsequently inhavitated the Royal Forests of the Eden Valley until those were 'disforested' in the sixteenth century. They are thus unique for, although red deer herds occur in other parts of England, those are not of unquestioned indigenous stock. There are also red deer in Grizedale and Thirlmere Forests and elsewhere in the Lake District, but they are extremely secretive and shy as the result of harassment and persecution (Plate IV).

The broken terrain, copses, caves and old mine workings of the fells provide an ideal refuge for the Fox. It is claimed that Lakeland animals are lankier and more tawny than their lowland counterparts and lack the black 'socks' of the latter. However, the existence of a distinct race of 'hill' foxes is a subject of controversy.

More remote and inaccessible crags provide nesting sites for three of the most interesting Lakeland birds, the Peregrine, the Raven and the Buzzard. The Peregine is the largest, boldest and most dashing of the British falcons. For half a century the population of this bird remained remarkably stable despite the attention of egg-collectors, falconers, gamekeepers, and increasing disturbance by rock climbers and others seeking recreation in wild country. In the early 1960s, however, a dramatic decline in its numbers occurred with an almost complete failure to breed. Chemical analysis revealed high concentrations of pesticide residues in the tissues of dead birds and in eggs that had failed to hatch. Further investigation showed that where the concentration of these substances was sub-lethal, birds either failed to nest or to lay, or if they did the eggs were infertile or so thin-shelled that they were inevitably broken when brooding was attempted. In due course a ban was placed on the use of aldrin, dieldrin and heptachlor, the most toxic of the pesticides then in use, since when there has been a progressive, if slow, recovery of the population. The Lake District has been more fortunate than many other parts of the country in this respect, hence its importance in conserving the peregrine.

The Raven and the Buzzard were formerly widely distributed in Britain and, neither being averse to scavenging, in medieval times appear to have been as familiar a feature of town and hamlet as of the countryside. However, the changing pattern of agriculture, improved standards of social hygiene, and, not least, persistent persecution resulted in their being eliminated from all but the most remote parts of the north and west. In the Lake District they were able to find refuge and safe nesting sites on the more inaccessible crags and sea-cliffs. This habit of crag nesting has persisted, although some nests are made in trees, as presumably were the majority when the species was more widely distributed. In common with other predators and carrion feeders, pesticide residues were found to have accumulated in the bodies of ravens in the early 1960s; fortunately they did not occasion the same degree of mortality and disturbance of reproduction as in the peregrine.

The buzzard has suffered more at the hands of man than probably any other raptorial bird in Britain. Its large size, soaring habit and 'mewing' call invite the attention of all, whilst its ponderous flight and willingness to investigate and eat carrion render it particularly vulnerable to shooting, trapping or poisoning. Happily, a more tolerant attitude towards this bird has prevailed recently so that from being reduced to a few breeding pairs it has recovered much lost ground. In 1955 the virtual disappearance of the rabbit, following

PLATE V. Castlerigg stone circle, set in dramatic scenery, is of the late Neolithic period

PLATE VI. Sheep farming—a major contribution to the economy and the appearance of the Lake District

PLATE VII. An example of Victorian architecture, Windermere Hotel was opened to serve the newly arrived railway

PLATE VIII. Comb Gill corn and saw mill, in Borrowdale, is one of the many old-established water-powered mills

the advent of myxomatosis, induced a temporary decline in the Lakeland buzzard population; but numbers rapidly recovered, presumably because, in this area, sheep and other carrion had always comprised a significant component of the buzzard's diet. In the absence of rabbit, long believed to be the staple food, it simply increased its intake of carrion. Immediately another hazard presented itself in the shape of food, now the main supply, contaminated with pesticide residues. This resulted from the use of sheep-dips containing aldrin and other persistent pesticides and the accumulation of residues in the fleece and subcutaneous fat of the sheep. But no substantial change was noticed in the buzzard population before the general use of these pesticides was banned. The bird is now fairly widely distributed in the Lake District and many woodland sites have been colonised.

In the late 1960s an event occurred of outstanding interest in the ornithological history of the Lake District. The Golden Eagle returned to breed for the first time since the end of the eighteenth century. Birds first began to be reported as regular visitors in the early 1950s, at first in the winter, later during the breeding season; but not until the end of the decade were nests built, and another ten years elapsed before egg laying was finally confirmed. Regrettably the first clutch was deserted, almost certainly because of human interference and disturbance. The following year, with careful wardening, young were successfully reared. This has been repeated in the last few years, and one hopes that this majestic bird will soon become a permanent feature of the National Park. However, in such a small area, with so large a concentration of holidaymakers, its status must remain precarious and the wardening of nesting sites will be necessary.

The Mountain Ringlet Butterfly is perhaps the most noteworthy insect of the fells. It occurs at about 1,750 feet and extends well into the montane zone. It is not recorded from England and Wales outside the Lake District, where its traditional locations are Red Screes, near Kirkstone Pass, and the Langdales. By contrast the larger and more spectacular Emperor and Northern Eggar Moths are characteristic of the lower fells, especially areas dominated by heather. As both species are strong fliers they, or their caterpillars, may be encountered wherever heather and other heath plants occur. Unlike most moths, the males of these two species are active during daylight and are therefore more likely to be noticed by visitors; the emperor in particular, because of its light ground colour and conspicuous eye-spot on each wing. The large caterpillar of the emperor, about 2 inches long, is bright green in colour with black markings,

FIGURE 5

Butterflies and moths

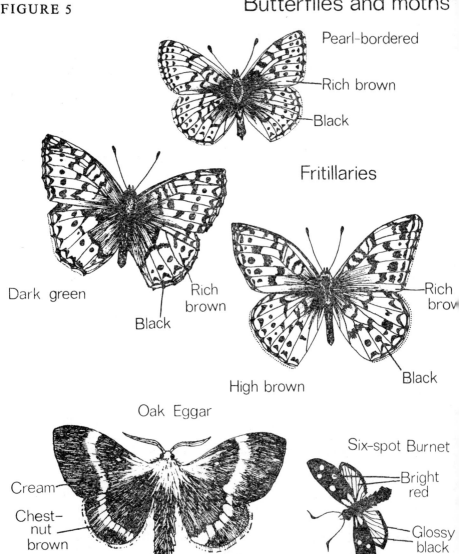

Pearl-bordered — Rich brown — Black

Fritillaries

Dark green — Rich brown — Black

High brown — Rich brown — Black

Oak Eggar — Cream — Chestnut brown

Six-spot Burnet — Bright red — Glossy black

Large Heath — Orange/brown

Grayling — Light orange — Dark brown

whilst that of the the northern eggar, at up to $2\frac{1}{2}$ inches in length, represents the ultimate in outsize 'hairy' caterpillars. (See also Figure 5.)

A bird characteristic of the lower fells, especially where heather is extensive, is the Merlin. This small falcon is frequently mistaken for the Kestrel. It can be readily recognised by its habit of perching on boulders, low walls and fences, and by its swift purposeful flight, usually fairly close to the ground. These features are in contrast to the kestrel's preference for higher perches such as telegraph poles, trees and buildings, and its drifting flight between hovering stations. In general the kestrel is more characteristic of lowland country, although it is not uncommon on the fells and nests have been found up to 1,700 feet. Regrettably the Red Grouse, a species or race of bird found nowhere outside Britain, is poorly represented in the Lake District. Small numbers occur in the northern part of the National Park and occasionally on some of the Cumberland mosses; but nowhere do population densities compare with those of the Pennines.

The characteristic habitat for the Nightjar is where heather moor merges into open woodland. The dearth of such transitions might therefore be expected to result in the absence of this bird in the Lake District. In fact it appears to find the transition from woodland to moss equally attractive, so that it is not uncommon, particularly in the southern part of the National Park. It is, however, more likely to be heard than seen, usually at dusk. Its churring 'song' is reminiscent of a distant two-stroke engine.

Many other species occur on the lower fells, for example, Curlew, Lapwing, Golden Plover, Snipe, and, along water-courses, Sandpiper, although none of these is so abundant as in the more open country of the Pennines. Other species associated with fell streams are Grey Wagtail, Yellow Wagtail and Dipper, but the last two occur more commonly at lower altitudes.

The woodlands. Woodlands may be classified as broadleaved, coniferous, or mixed, *i.e.* mixtures of broadleaved and coniferous trees. The broadleaved woods include all that remains of the original native forest, although much modified as the result of a long history of human interference. The coniferous woodlands, with the exception of some self-sown pinewoods, as on Rusland Moss, have been planted comparatively recently.

The highest broadleaved woodlands are the Keskadale and Birkrigg oakwoods at 1,400 and 1,250 feet respectively. In these the trees are stunted and gnarled—in parts of Birkrigg it is possible to walk 'through the woods' with one's head above the leaf canopy. Oak is often the only tree species in these woods and, as might be expected, they support only a sparse population of animals. The commonest breeding bird is the Chaffinch; other species are Pied Flycatcher, Coal Tit, Redstart, Wren, Willow and other Warblers, and the ubiquitous Robin.

The lower and more extensive broadleaved woods are correspondingly richer in animals. All three British species of Woodpecker occur, Green, Greater Spotted and small Lesser Spotted, although the last is frequently overlooked because of its habit of keeping to the tops of trees and consorting in the winter with flocks of tits, etc., among which it is not readily noticed. Several of the woods are well known, especially to sportsmen, for their winter populations of Woodcock. This species is also resident and breeds in most suitable woodlands. It is extremely secretive during daylight and is usually encountered accidentally when it suddenly erupts at one's feet and flies off weaving its way between tree stems. During the breeding season the woodcock is most reliably encountered at dusk, when with slow moth-like flight it patrols its territory, usually the margin of a wood, emitting characteristic grunts and pipings.

The Sparrowhawk is now common in the Lake District. In the early 1960s the British population was decimated, and whilst numbers here were inevitably reduced, this seems to have happened on a lesser scale than in the country generally. This presumably was due to the limited amount of arable farming in the area and the lower risk of small birds, on which the sparrowhawk preys, picking up 'dressed' seed or invertebrates contaminated with persistent pesticides. At such a critical time, the relatively small amount of game-rearing in the Lake District probably also helped. Visitors cannot expect to see this bird as a matter of course. Except when coming to the nest to feed its young, it is extremely elusive and is normally seen only for a fleeting moment when it loops from the edge of a wood or slips over a hedge in search of prey.

The extensive planting of conifer woodlands during the last 50 years has had a profound effect on the fauna of the Lake District although, on a unit area basis, they are significantly poorer in species than deciduous or mixed woods. Exceptions are the young plantations which increasingly provide a habitat for woodland species yet continue to support those inhabiting the grassland or heathland before afforestation. Thus, birds of open country such as Partridge, Red Grouse, Meadow Pipit and Skylark may be found in association with woodland species such as Black Grouse, Pheasant, Chaffinch and Warblers. In addition, the exclusion of sheep-grazing and the shelter provided by growing trees results in an increased luxuriance of ground vegetation. This abundance of food and cover promotes an increase in the population of voles and mice, which in turn attracts predators. The most characteristic of these is perhaps the Short-eared Owl. This species is reasonably easily recognised as, with the exception of the Barn Owl, it is the only large owl likely to be actively hunting during daylight and its long narrow wings are a useful aid to identification. As the forest matures and the canopy closes, the ground vegetation is suppressed and the initial richness and variety of animals declines. That is ultimately reflected in the silence at the heart of a mature conifer plantation where, apart from the cooing of nesting Woodpigeons or the occasional screech of a Jay or the chatter of a Magpie, little birdsong is to be heard. The Chaffinch, as in the oakwoods, is the commonest breeding species of the conifer plantations, but the most characteristic are Coal Tit and Goldcrest. The latter is not easy to spot since it is our smallest bird, frequenting the outermost parts of the canopy, in which it also nests. Moreover when alarmed it does not create the shindig of the Wren, our next smallest breeding bird. Another characteristic bird of conifer plantations is the Crossbill. Although it still appears to be represented only by a small resident population in the Tarn Hows area, periodic winter 'invasions' of Continental birds take place. As its name suggests, the mandibles of this species are crossed, giving an effective means of prising open the scales of cones in order to extract the seeds. It frequents the outer branches of conifers, so that like the goldcrest it is not always noticed.

Although conifer plantations are not rich in animals on a unit area basis, they do cover large areas and thus provide effective refuges for species hard-pressed elsewhere. This has been particularly true for the Roe Deer. Formerly widespread in Britain, the roe was brought to the verge of extinction by deforestation and persecution. Small populations did survive south of the Scottish Border, in the Lake District, north Northumberland, and one or two other areas. The plantations of conifers in the Lake District undoubtedly saved the native population from annihilation. Latterly a more tolerant attitude on the part of the foresters has resulted in its recovering lost ground to such an extent that only the smallest woodlands are now without a roe deer family. This does not mean that this graceful little animal will be seen by many visitors. It is extremely shy, adept at finding secluded spots in which to lie up and, except in the most undisturbed areas, predominantly nocturnal.

Another animal which has benefited from increased afforestation, and a more tolerant attitude to wild life on the part of landowners and the public generally, is the Badger. At the beginning of this century the badger was undoubtedly a much rarer animal than it is today. It eats both vegetable and animal food including berries, grass roots, bulbs, earthworms, beetles, young rabbits and carrion. Its sett although characteristically located in woodland may be found on the open fell, in hedgerows or river embankments, caves and mineral workings, or on rubbish tips. A most adaptable animal, its rarity in the past can have arisen only as the result of persecution in ignorance of its habits or because of misguided tradition and superstition. Although it is not especially attracted to extensive plantations of conifers, within which little food may be available, these undoubtedly provided a refuge in the recent past from which animals were able to move out to colonise more favourable habitats as opportunity arose. Like the roe deer, the badger is nocturnal and thus rarely seen, unless it happens to be picked up in the headlights of a car or becomes a road casualty. It is not difficult to watch by those who have the patience to sit quietly and still near its sett at dusk.

The Pine Marten, one of the rarest British mammals, still survives in the Lake District. It is related to the otter, badger, polecat and stoat but is more arboreal than any of these. It appears to prefer an open woodland habitat which may be pinewood in Scotland or broadleaved or mixed woodland in Wales and the Lake District. It is said that the species survived in Lakeland by retreating to the central fells; however, records in the southern part of the district during the last 50 years usually relate to lowland sites such as Rusland Moss and the Newby Bridge area. The state forests have undoubtedly provided a refuge if not an ideal habitat for this species. The pine marten is a ruthless killer when it gains entry to coops of chickens or pheasants, so it has inevitably suffered at the hands of farmers and keepers. The tradition of keeping chickens on every smallholding has declined in recent years, and where it continues, birds now tend to be housed in purpose-built predator-proof buildings rather than in the makeshift dilapidated structures that formerly often served this purpose and offered an open invitation to enter to every hungry predator. So there should be less cause for landowners to take action against the pine marten; indeed it could prove a useful ally to forestry interests by helping to control the numbers of grey squirrels and woodpigeons that inevitably breed in extensive plantations in the absence of natural predators.

The Polecat, nearest relative of the pine marten, is periodically reported from the Lake District. It is, however, difficult to distinguish from the coloured form of the domestic Ferret, so that one cannot be certain to which species recent sightings relate. The situation is confused by the fact that, with the advent of myxomatosis, ferrets are known to have been released 'to the wild' to fend for themselves. The polecat was formerly a widely distributed species in England and Wales, but as the result of persecution was eliminated from all parts of its range except Mid-Wales. There its habitat included sand-dune, bog and woodlands to over 1,000 feet. In the main it appears to be more of a lowland animal than the pine marten, and there is obviously an abundance of habitats in the Lake District suited to its needs. In recent years the polecat has been recovering lost ground in Wales to the extent that it has been recorded from most counties; there is therefore just a possibility that some of the recent Lake District reports relate to the expansion of a small native population that managed to survive somewhere in the area, or even to animals that have recolonised it from elsewhere. Confirmation of its presence would be gratifying not only as establishing a complement of native Lakeland mammals but also in ensuring a better balance of natural predators.

The status of one mammal, the Red Squirrel, gives cause for concern. Formerly widespread, this species has become increasingly restricted in its distribution. The reasons for its decline are not fully understood. One is often told that it has been ousted from its former habitats by the Grey Squirrel, a species introduced from North America. There is, however, evidence that red squirrel populations often decline before the appearance of greys, and this is certainly true of the Lake District. One thing is clear: where the red squirrel is supplanted by the grey it appears to be virtually incapable of re-establishment. Although the red squirrel seemed to be reaching a peak of abundance in the Lake District about 1970, it is likely that this was no more than a prelude to one of the cyclic crashes to which the species is subject. Since the grey squirrel has now reached the southern part of the district it may be only a matter of time before the red squirrel disappears as a member of the fauna.

The invertebrate fauna of the Lake District woodlands is legion, ranging from the attractive Orange-tip and Brimstone Butterflies to a myriad of inconspicuous creatures associated with trees and their by-products.

The waters. The lakes, tarns, ponds, rivers, streams and ditches of the Lake District provide a great diversity of fresh-water habitats. A vast population of animals, largely unseen, lives below the surface of the water (Figure 6). It comprises molluscs, crustaceans, insect larvae and a host of microscopic creatures, as well as the more

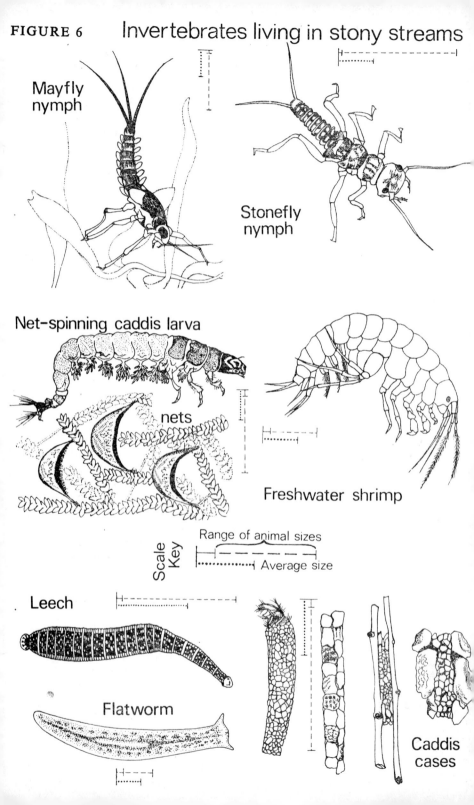

Animals of the National Park

familiar fishes. Whilst many of these waters are famous for their salmon and trout, perhaps the most noteworthy fish is the Char. Izaak Walton believed that this species was confined to Windermere, but this is not so. Especially associated with that lake, however, is the Netted Carpet Moth which occurs in some of the woodland bordering Windermere and nowhere else in Britain. (Incidentally *Lysimachia terrestris*—a plant of the loosestrife family—has a similar distribution.)

The lakes necessarily provide the most significant fresh-water habitat of the Lake District. Some have changed little since the last glaciation, but others, such as Esthwaite and Bassenthwaite, have been much modified by the accumulation of inorganic and organic debris creating rich habitats for animals and plants. The more evolved lakes, with their shallow margins supporting reedswamp and other emergent vegetation, sheltered embayments and islands, provide a habitat *par excellence* for aquatic birds. Esthwaite, considered to be the most highly evolved of all, supports a stable population of Great Crested Grebe, and the species nests fairly regularly on other lakes and can be seen on them outside the breeding season. Several species of duck, Golden-eye, Pochard, Wigeon, Shoveller, Tufted, Teal and Mallard, utilise the lakes as winter feeding grounds and roosts, although only the last-named species occurs in really large numbers and only the last three breed regularly. The fish-eating sawbills, Merganser and Goosander, can also now be regularly encountered. The former first appeared as a breeding species at Ravenglass Gullery soon after this was established as a Nature Reserve by Cumberland County Council, since when Windermere and other waters have been colonised. Although the goosander has not yet established itself as a breeding species in the National Park, numbers occur as winter visitors to Ullswater and Coniston Water in particular. All three Divers, Great Northern, Black-Throated and Red-Throated, may be seen as occasional, if somewhat irregular, winter visitors.

The most conspicuous winter visitors to tarns and lakes alike are the Whooper and Bewick's Swans. Family parties of the former may be encountered on even the smallest tarns, where they rapidly exhaust the food supply. In severe weather herds of two or three dozen birds may congregate on the larger, more productive lakes. The smaller Bewick's swan is a much less regular visitor whose numbers rarely compare with those of the whooper. Both can be distinguished from the commoner, and tamer, Mute Swan by their habit of sailing with the neck held straight, rather than in an S-bend, and by their yellow, rather than orange, bills.

Gull roosts are a winter feature of several lakes. Most noteworthy is Ullswater with its thousands of Common Gulls, but also Black-headed and, increasingly, Herring and Lesser Black-backed Gulls. The most characteristic and possibly most abundant winter inhabitant of the lakes is the Coot, of which vast rafts may assemble on Windermere and other of the more productive waters.

The carrs, reedswamps and marshland associated with fresh waters provide rich feeding grounds and nesting sites not only for aquatic birds but for a wide variety of other species such as Reed Bunting, Sedge and Grasshopper Warblers, waders such as Snipe and Sandpiper, as well as miscellaneous individuals like Pheasant and Corncrake. The latter is now, as in Britian generally, extremely rare but still occurs in one or two places in the south of the National Park where rank marshy vegetation cannot be mown mechanically or is too poor for haymaking.

The Heron is unlikely to pass unnoticed, because of its large wing-span and its raucous croak of alarm, as it springs from lake or tarn margin, river, stream, or even damp pasture, where frogs and other foods are to be found. Despite the loss of several long-established heronries as the result of tree-felling, the species has managed to maintain its numbers remarkably well. There are still one or two small heronries in the National Park, particularly the southern part, but the population of Lake District birds is almost certainly maintained by recruits from the large heronries outside its boundaries such as the historic one at Dallam Towers, Milnthorpe. Fish-eating birds like the heron are particularly vulnerable to pesticide poisoning, for not only are those substances used to clear waterways of vegetative growth but persistent pesticides applied to the land ultimately drain into the water-courses. As with other species previously mentioned, the heron population of the Lake District has fortunately suffered less from such effects than those of many other parts of the country.

The recent history of the Kingfisher, another fish-eater, has been less happy. The Lake District population appears to have been declining for many years and since the early 1960s its numbers have been very low. Although this trend coincides with the period when toxic chemicals were causing the greatest havoc, it is probably more closely related to high mortality in the severe winter of 1962–63 when many of the streams, rivers and lake margins were frozen for long periods, isolating birds from their food supply.

The Otter is undoubtedly the mammal most likely to be associated with fresh water in the mind of the visitor, although this is equally true of the Water Vole (locally and inaccurately referred to as the

Water Rat) and the Water Shrew, both of which are infinitely more common and widespread. The status of the otter is always difficult to determine and this is particularly so in the Lake District with its exceptionally high proportion of suitable habitat. Regrettably such information as is available suggests that this attractive and playful creature is becoming increasingly rare in the National Park. This is probably due to a complex of factors: reclamation of wetlands, increased disturbance of former sanctuaries by those seeking recreation in them, otter hunting, and possibly, by no means of least importance, the effects of pesticides and pollution on food resources. Of particular concern are the astronomical rents now being paid for salmon and trout fishing. It would be taxing credulity to the limit to suppose that this does not place on water-bailiffs an undue pressure to ensure that anything militating against a full return on outlay, such as predation of fish by the otter, is rapidly eliminated. The future looks black since the survival of the otter is dependent on so many interests.

The coastal area. One of the most interesting places is Ravenglass Dunes, established as a Nature Reserve by Cumberland County Council in 1954. This supports the largest breeding colony of Black-headed Gulls in England, believed to represent about one-third of the total population. Sandwich, Common, Arctic and Little Tern also breed here: the last-named, as in the country generally, somewhat precariously. On the establishment of the Nature Reserve the numbers of Sandwich Tern rapidly increased to about 500 breeding pairs, a tenfold increase on post-war numbers. Other breeding birds are Ring Plover, Shelduck and Oystercatcher.

West of the National Park, St. Bees Head is noteworthy for its cliff-nesting Puffins, its colony of Fulmars and one or two pairs of Black Guillemots. It is also a traditional stronghold of the Stonechat, a declining species generally. Similarly, Walney Island to the south supports the largest breeding colony in the district of Herring and Lesser Black-backed Gulls and provides the southernmost breeding station in Britain of the Eider Duck. The Roseate Tern used to nest here regularly, if in small numbers, until the loss of habitat by gravel-working and human disturbance resulted in its forsaking the area in favour of some of the adjacent coastline.

The mosses because of draining are much overgrown with scrub and self-sown woodland so that their animal life is now more characteristic of acid woodland than open moss. Until the beginning of the last war a large colony of lesser black-backed and herring gulls bred on Foulshaw Moss. However, the collecting of eggs to supplement supplies of dried-egg powder was apparently too much for the colony which, by the end of the war, had totally collapsed. Changing character of the mosses also undoubtedly accounts for the disappearance of the pre-war breeding population of red grouse. On the other hand, the drier condition of the mosses and the increased cover they now provide has been a boon to roe deer, hares, foxes and many bird species previously mentioned. They are also the most profitable place in the National Park to look for reptiles. Adder and Common Lizard are abundant whilst Grass Snake and Slow Worm are locally common. The Sand Lizard, as its name indicates, is more associated with coastal dunes than inland country. Although occurring in Lancashire, it has not yet been recorded 'North of the Sands'. This is in contrast with an amphibian, the Natterjack Toad. recently reported from the Ravenglass area; previously its nearest known location had been the Southport dunes. The Common Toad is found throughout the area although not at such high altitudes as the Frog. The latter, which appears to be a declining species in England, possibly as the result of the contamination of its food by persistent pesticides, is still reasonably common in the Lake District and may be found almost up to the highest tops. Three other amphibians, Great Crested, Palmate and Common Newts, tend to be local but where they occur may be quite abundant.

No account such as this would be complete without reference to Morecambe Bay. As the result of recent surveys it was established unquestionably that the Bay is the most important area in the whole of Britain and possibly N.W. Europe for the conservation of wintering waders. It is also a staging post for waders passing through the area on spring and autumn migration. The large flocks of Dunlin (Leven estuary), Knot (Kent/Keer estuary) and Oystercatcher are a spectacle which visitors should make a point of seeing, because if some of the proposals to store water in the Bay are proceeded with the experience may be something which they will only be able to *tell* their children or grandchildren about. The Leven estuary is a particularly good place to observe some of the two to three thousand Shelduck of the Bay—a figure that qualifies it as a site of national importance for this species. The Kent estuary is of special importance for its population of Greylag Geese. These geese have been regular winter visitors since soon after the First World War; up to about 200 appear each year. But there is evidence that the range of this species is contracting (most do not now come farther south than Scotland), and this may soon become the only place in England and Wales where it can be regularly seen. (Birds seen on inland tarns during the summer have been set down by local wildfowlers.)

Animals of the National Park

Finally the Bay is well known for its Shrimps and Flukes (Flounders) to the extent that Flookburgh has been named after this fish and its village church boasts a unique weathervane in the form of it. For those who wish to learn more about the variety of wild life of the Lake District, the short Bibliography on page 139 will go some way. In the last analysis, however, the best and certainly the most interesting way to become familiar with the fauna is to observe it in the field; taking care of course not to disturb the animals, and plants, in the process. Much still remains to be discovered about the lives and populations of even the commonest species, so that such observations, in addition to being an agreeable pursuit, may well augment our knowledge.

5
Remains left by Ancient Peoples

by CLARE I. FELL

ALTHOUGH many parts of Cumbria which were favourable to early settlement lie outside the artificial boundaries of the Lake District National Park, remains of people who lived here in the distant past do occur within its confines. Many of the surviving field monuments, dating from prehistoric times until the Norman Conquest, are protected under the Ancient Monuments Acts; these are listed in Appendix I. In addition, small finds, giving some insight into the material culture of these ancient peoples, can be seen in various museums, especially at Carlisle (Tullie House, Castle Street) and Lancaster (Old Town Hall, Market Place). A few finds are also exhibited at Fitz Park Museum, Keswick; Ruskin Museum, Coniston; Barrow-in-Furness Museum (Ramsden Square); and Kendal Museum (Station Road).

Prehistoric sites and finds. Surface finds of flints of the Mesolithic period provide the earliest trace of man in this district. These have been found so far only along the coastal strip, where forest growth, which developed after the end of the Quaternary Ice Ages, is likely to have been less dense. Worked flints include microliths for mounting in bone and antler shafts as arrow-tips, barbs for harpoons, and other equipment used by the hunting communities of that time. Such flints have been found at St. Bees and Drigg outside the Park, and within it at Eskmeals. No firm dating is yet available though it is possible that the flints were fashioned between five and seven thousand years ago.

With the development of agriculture and other skills in the Neolithic period, roughly from the fourth to the early second millennium B.C., there is evidence that parts of the district were extensively exploited. Numerous polished and roughed-out stone axes, needed for forest clearance, and occasional finds of smaller flint tools and pottery characteristic of that time have survived. Few settlements or burial sites are yet known and none have been excavated by modern methods. The most interesting habitation site was discovered last century at Ehenside Tarn, near Beckermet, when the tarn was drained for agricultural purposes. Pottery, stone axes in many stages of manufacture, a saucer-quern and grain rubbers, and items of wooden equipment were found at hearths around the shore. Conventional radio-carbon dates, ranging from 3014 to 1570 B.C., have been obtained from several wooden objects from the site. More recently evidence of Neolithic settlement has come from Williamson's Moss, Eskmeals, while finds at Mossgarth, Portinscale near Keswick, early this century give evidence of stone-axe manufacture.

There are a number of possible Neolithic long cairns within the National Park, though the only one certainly accepted as a burial site of that period is on Stockdale Moor, half a mile north of the river Bleng beyond Gosforth. It is known as Sampson's Bratful and legend holds that this stone heap fell from the devil's apron (*brat* means apron in Old Welsh). Many other cairns of Bronze Age character can be seen on the same moor.

Perhaps the most important discovery in the central hills of the Lake District has been the Neolithic stone-axe factories, first noticed on Mart Crag Moor, near Stake Pass, at the head of Great Langdale, then at various sites on Langdale Pikes and more recently in the Sca Fell–Scafell Pike–Glaramara region. The rock used was a fine-grained, greenish-grey tuff of the Borrowdale Volcanic Group; the products were widely traded throughout Britain. There is as yet no evidence that the tools were finished at the factory sites, and the polishing process seems to have been done at settlements on the coastal fringe and in other areas suitable for primitive agriculture, especially in the limestone districts. Recently two conventional radio-carbon dates have been obtained from charcoal found among chippings on the north side of the Pikes, giving 2730 and 2524 B.C. for one phase of the workings.

The most striking survivals from prehistoric times here are the large stone circles and their counterparts the smaller cairn-circles, which surround burials (Figure 7). Much has been written recently about the astronomical use and geometric construction of the large circles, but whatever the truth of this may be it is certain they were of great ceremonial significance to their builders in the late Neolithic or early Bronze Age times. Elsewhere in Britain such circles are sometimes connected with the Beaker people, who spread into the country from the Continent in the centuries around 2000 B.C. and who had some knowledge of the use of metals. The circles within the Park remain undated and few traces of the Beaker culture have been found, though it is represented by a number of burials in the Eden Valley. Castlerigg, near Keswick, is set among splendid hills

FIGURE 7

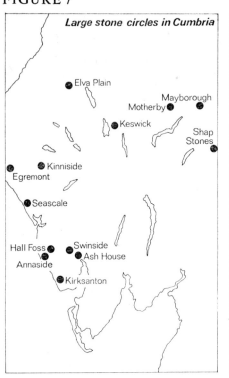
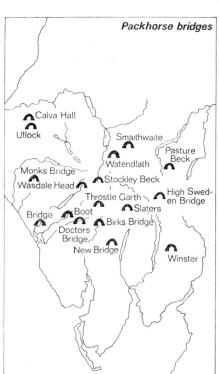
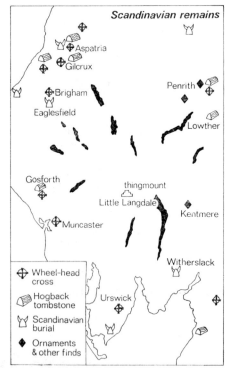
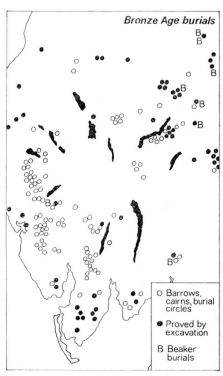

Remains left by Ancient Peoples

and has an unusual rectangular arrangement of stones inside its perimeter (Plate V). The stone circle known as Swinside, or Sunkenkirk, about 5 miles north of Millom is also finely sited, and there is another damaged circle at Elva Plain, east of Cockermouth. All are in areas where finds suggest a considerable population existing at that time.

The smaller Bronze Age burial circles vary in construction. In some, free-standing stones encircle a central cairn, as at Standing Stones, Moor Divock, Askham; others have a bank of stones and earth enclosing an area in which cremation burials, often in urns, have been deposited, as on Banniside Moor, Coniston; while on Burn Moor above Eskdale, five burial cairns are enclosed within a circle of standing stones. There are many scattered burial cairns as well as cairnfields (*i.e.* large groups of small stone heaps which may not all contain burials, but may indicate field clearance from some vanished system of agriculture). These can be seen on Thwaites Fell, on Bootle and Corney Fells, on the hillsides north of Carrock Fell, and on Heathwaite and Woodland Fells in Furness, to mention but a few examples. Isolated barrows on hilltops can also be seen, such as those on Great and Little Mell Fell west of Dacre; the few which have been excavated have yielded Bronze Age urns.

Bronze tools and weapons of all phases of the Bronze Age (*circa* 1700–500 B.C.) have been found from time to time within the National Park. These occur not only on the lower ground along the coast and in the Keswick, Vale of St John, and Ambleside environs, but also in the hills, for instance on Woundale Common above Troutbeck (Windermere), on Low Fell, Little Langdale, at Dalehead, Martindale, and near the fort up Shoulthwaite Gill above Thirlmere.

Apart from flint-working sites in the coastal sandhills, settlements of the Bronze Age are not as yet known, though it may be that some of the hill-forts, once thought to date exclusively from the pre-Roman Iron Age, could have been first established in the later part of the Bronze Age. The largest of these, about 5 acres in extent, is on the summit of Carrock Fell and is surrounded by a single fallen stone rampart. No traces of circular stone-based huts are visible within it; buildings which formerly existed there are likely to have been made of timber. Other forts inside the National Park are at Shoulthwaite Gill at the north-west end of Thirlmere, Castle Crag, Mardale, and Castle Crag, Borrowdale. The last has produced the only dating evidence so far available in the form of Roman pottery, including imported Samian ware and other Roman material. This suggests trade with, or looting from, the Roman forts by native tribesmen in the early part of the occupation. The other forts are not yet dated and few settlements of the pre-Roman Iron Age are known.

Celtic-style metalwork is occasionally found within the Park; for instance, a pair of ribbed bronze bracelets were picked up below Rough Crag, Thirlmere, and a bronze bridle-bit from screes on Place Fell, east of Ullswater. A fine iron sword with bronze and enamel hilt-mounts and scabbard was found in the course of draining and enclosing Embleton High Common early last century and shows Roman influence on native Celtic metalwork.

The Roman occupation. Military occupation of the north-west began in the years following the suppression of the revolt of the Brigantes tribe in A.D. 69 and continued until late in the fourth century. A few Roman roads and auxiliary forts lie within the National Park, though the bulk of the defensive works and arterial roads lie around the Lake District hills. The fort and supply port of Ravenglass (*Glannaventa*) is much damaged by the railway, but its bath-house, known as Walls Castle, still stands to a considerable height. From Ravenglass a road led eastward up Eskdale to Hardknott fort (*Mediobogdum*), now in the guardianship of the Department of the Environment. This fort was built early in the second century. Its parade ground, levelled out of the hillside above, can still clearly be seen. From Hardknott fort the Roman road continuing eastward climbed over the pass, followed the flanks of Wrynose Bottom, over Wrynose Pass, and so along Little Langdale to another fort at the head of Windermere, Borrans Field (*Galava*) at Ambleside. One connecting link with the main arterial road climbs over High Street on its way to Brougham (*Brocavum*), while another branch, not fully traced, probably continued south-east to Watercrook near Kendal and from there to Burrow in Lonsdale. Another Roman road struck westward from Old Penrith (*Voreda*) and traces of four Roman fortifications are known near Troutbeck railway station, now disused, about 9 miles east of Keswick. North of Bassenthwaite Lake a temporary camp and later fort and fortlet can be seen at Caermote, though little is yet known of its history.

The farmsteads of the native population survive in many valleys and hillsides of the Lake District, for example, in Bannisdale, Mardale; Glencoyne Park, and Grisedale on the west side of Ullswater. They are not so well known as the group in the limestone country to the east of the National Park around Shap, Crosby Ravensworth, Crosby Garrett, and Waitby. Most of these farms appear as irregular enclosures, bounded by massive walls 5 feet or more wide, faced with large boulders on both sides and a rubble-filled core. They mostly contain circular stone-based huts and irregular pens for stock, the walls often being grass-grown with the passage of time. One or more entrances give access to these settlements which were lived in by a predominantly pastoral people. Many cover less than half an acre and are likely to have been worked by single families, but there are others more than twice that size which suggest a small village, such as the homesteads up Heck Beck in Bannerdale and those at Hugill near Windermere and at Millrigg and Tongue House in Kentmere. Smaller farms can be seen, for instance, on Skirsgill Hill, Askham; east of Torver Beck, Banniside Moor, Coniston; at Whitrow Beck in Waberthwaite parish; and at Lanthwaite Green near the foot of Crummock Water. Three remarkable enclosures on Aughertree Fell, Ireby, are worth a visit.

After the Romans. How long these farms were occupied after the end of the fourth century is not known, nor whether some of the more accessible fortified sites, such as Castle How, Peel Wyke, Bassenthwaite, or Dunmallet, Pooley Bridge, at the foot of Ullswater, were constructed or reoccupied at that time. Nor do we know how long the legacy of administration built up under the Romans survived in the British kingdoms which succeeded to their authority. From place-name evidence, now available for both Cumberland and Westmorland in the volumes published by the English Place-Name Society, it is known that the Anglian kingdom of Northumbria dominated much of this area from about the middle of the seventh century until the Viking settlement of the early tenth century. This latest large movement of people into the district greatly affected the central hills, for the Vikings were predominantly pastoralists, rearing cattle, pigs, and sheep, and were not so dependent on the richer agricultural land suited to the arable farming of the Anglians.

Field monuments of both these peoples are scanty. A few pagan Anglian burials in earlier barrows have been found at the head of the Eden Valley but none in the National Park. A Viking sword, probably from a man's grave, was found on Whitbarrow Scar near Witherslack, but non-Christian Viking burials in Cumbria have otherwise been found outside the Park. Houses of both these peoples have not been identified with certainty because many will have been built of turf or timber and overlaid by later stone buildings, or their foundations destroyed by ploughing and other development. The drystone walls of rectangular shieling huts at Old Whelter, Mardale, are thought by some, on place-name evidence, to date back to the eleventh century. However, no datable finds were made during excavations there in 1922. The system of pasturing stock on the open fells during summer at some distance from more permanent farms and enclosed fields continued in the Border region down to the end of the seventeenth century.

Anglian metalwork and coin hoards from Cumbria are also scarce and have not been recorded from places within the National Park; the same is true of Viking period finds, apart from an iron spear-head from Kentmere, probably dating from the tenth century. Of similar date, from just outside the Park, are two fine, silver 'thistle' brooches, one from Fluskew Pike north of Dacre and the other, not closely provenanced, 'found near Penrith'.

The finest surviving monuments of Anglian and Viking times are the sculptured stone crosses and hogback tombstones. Most of these are preserved at churches outside the Park, but a fine Anglian cross can be seen in Irton churchyard and another at Waberthwaite. Fragments of two Anglian crosses, one known as 'the lion stone', are preserved at Dacre church, where also are the probable foundations of a pre-Conquest monastery. The most famous Viking wheel-headed cross, probably of early eleventh-century date, is at Gosforth, carved in red sandstone with scenes from Norse mythology as well as with Christian symbols. Inside the same church are further interesting cross-fragments and hogback stones, fashioned in the shape of houses of that period. Other Viking period sculpture, to be seen at Muncaster and Isel churches, points to the close link which existed between the Vikings of Cumbria and those of the Isle of Man and Ireland.

6
The Medieval Scene and After

by J. C. DICKINSON

THE primitive technology of medieval times together with the poor communications and inhospitable nature of the Lake Counties in general and the National Park area in particular made it inevitable that throughout the Middle Ages the district should be very thinly populated. No small part of the road system thrown across Lakeland by the indomitable Romans fell into disuse after the collapse of their rule, and no sizeable centres of population developed in the Park area until the railway brought tourists in force to Keswick, Ambleside and Windermere. The major medieval castles of Lakeland all stood on the fertile fringe which surrounds the Park, as did most of the medieval monasteries, though Calder Abbey is just within one side of it and Shap Abbey just within another. As late as the fifteenth century the population lived largely in villages, hamlets or isolated farmsteads.

For most of the medieval period literary sources for the history of the area are sparse, leaving large gaps in our knowledge, of which many are unlikely ever to be filled. From the sixth century onwards much of the area belonged for some time to the Celtic kingdom of Cumbria with its capital at Dumbarton on the Clyde. But its frontiers are uncertain and were not stable, the whole area for long after the Norman Conquest being something of a no-man's-land. During the sixth and seventh centuries not very numerous Anglian colonists pushed into the Park area from the Eden Valley and lands around Morecambe Bay. But they were people who liked good farming land so left much of the Park untouched, though amongst other things they gave their names to Keswick ('cheese farm'), Buttermere (the lake adjoined by pastures yielding plenty of butter) and Staveley ('the wood where staves were got'). Of the old Celtic people, who went on 'living and partly living' in the infertile central lands of the Park area, we know little, though they gave their present titles to not a few Lakeland rivers, such as the Calder, Cocker, Derwent, Eden and Irthing.

A new era opened in the history of the region with the arrival of the Vikings. These were mostly Norse and many of them came not direct from Scandinavia, but from Ireland and the Isle of Man, the earliest arrivals grounding their craft on Cumbrian beaches around the opening years of the tenth century. Main areas of the Viking landings were around St. Bees Head to the west of the Park and along the northern side of Morecambe Bay. The process of taking over the thinly populated land was gradual and not complete even by the twelfth century. Although some raids may have been accompanied by violence, there was plenty of unutilised land, and it was mainly as settlers that the Vikings left their mark. Place-names show that the National Park was predominantly Scandinavian in culture.

As W. G. Collingwood pointed out:

'If we try to recast the progress of Norse settlement in the Lake District, we find that the place-names are curiously suggestive. At the mouths of the valleys which converge towards the centre of the fells there is nearly always a place (often a -*by*) named after some early owner. We cannot date him in every case to the beginning of the settlement, but it looks as though he were the first to be remembered of the important family which lived at that spot. Then a little higher up the valley we have *booths* or smaller centres which often have Gaelic personal names as parts of their titles; and we know that the Norse settlers in Ireland took with them Irish or Scottish thralls and retainers whose names were given to the dependent farms they founded. Higher up are the summer shielings which grew into sheep farms. At the heads of the valleys there are nearly always names which show that the wild forest was used for keeping pigs.'

Various place-names incorporate names of Viking leaders like the Vinnunder of Windermere, the Thorstein of Thurstanwater (the old name for Coniston Water), the Asmundr of Osmotherley, and the Anundr of Ennerdale. The part of west Cumberland known as Copeland has a Norse name meaning 'the bought land', though the name of the original buyer and vendor are lost; the second element of Dunmail Raise derives from a Scandinavian word for a cairn which was apparently built there to commemorate Dunmail, the last king of Cumbria. Other Norse words are 'beck' (stream), 'gill' (a ravine), 'thwaite' (clearing or paddock) and 'wath' (ford). Of very few visible remains of the Viking age, much the most notable is the lofty Gosforth cross, carved with a curious mixture of pagan and Christian motifs about A.D. 1000 and easily the most spectacular thing of its kind in England. Local dialect remains heavily indebted to the Vikings, *e.g.* properly educated schoolboys ask 'wha's laikin', not 'who is playing'.

The famous Domesday Book of 1086 shows us that at that time the only parts of Lakeland surveyed as part of the English kingdom were the Kendal area of Westmorland, Lancashire north of the Sands—*i.e.* Furness and Cartmel—and a very small strip of Cumberland in the Millom district. But in 1092 William Rufus acquired

The Medieval Scene and After

Carlisle and over the next two centuries his successors sought tenaciously to maintain their hold on the area. To aid the process of Normanisation, a new bishopric was set up with its seat at Carlisle, the first bishop being consecrated in 1133. The struggle for control hereafter fluctuated violently for some time. By the mid-twelfth century the Scots had pushed their frontier south of Morecambe Bay, but that was a brief success. In 1172–3 they launched a big invasion of northern England with early victory followed by a disastrous defeat. The greater resources and sophistication of the English and their castles in the Eden Valley steadily strengthened their position, and by the middle of the thirteenth century their hold over Cumberland and Westmorland was definitely established, though throughout the Middle Ages and beyond the English were unable to hold off Scottish raids, some of alarming severity.

To control the land and its holders the Normans set up castles at centres outside the present Park (like Kendal, Appleby, Cockermouth and Egremont), but including in their baronies much land inside it.

In the heart of the Lake District, there was then probably little of value to defend and there are few signs of the early fortifications of those days. A little later defended dwelling-houses appeared, like the moated 'grange', a square or rectangular enclosure surrounded by a wide water-filled ditch and palisade. Some traces of one of these remain at Trostermount on Ullswater. Later, especially after the Scottish invasions of 1316 and 1322, the local gentry built peletowers, a form of local defence widespread in the Border counties. These were simple towers of stone designed to provide refuge until the raiders had gone their evil way. They had three or four floors. The bottom one was largely for stock and was entered by the smallest of doors; from it wall-stairs led to the living quarters above. The roof was generally designed for a beacon and there seems to have developed a careful system for signalling by these from one area to another. Of the few to be seen within the Park, that at Dacre is notable and unusually elaborate.

However, it must not be thought that the centuries after the Norman Conquest brought the Lake District nothing but woe. By the late twelfth century both Church life and economic life were expanding and were to continue to expand. The ancient kingdom of Cumbria was Christian at least in name and at an early date there were parish churches around the present National Park, and at least one in it, Crosthwaite church near Keswick, dedicated to St. Kentigern. When the Normans established their rule, the chief parishes, like the baronies, were enormous straggling affairs stretching into the mountains from the fertile lands outside. Thus in the Cartmel, Kent and Lune Valleys, we find the great mother churches of Kirkby Cartmel, Kirkby Kendal and Kirkby Lonsdale dominating a large but thinly populated area. (*Kirkby* = church town.) Such churches had originally almost a monopoly to baptise and bury those resident within their parishes, so that baptisms and burials in these sprawling areas were often inconvenient, especially in winter when bad weather might render the tracks that snaked along the fells almost impassable. As population increased, little chapels were built to serve local communities and at least make it possible to hear Mass easily. Some of these were in private houses, others founded by the locals as a co-operative effort. Grasmere and Bowness-on-Windermere are known to have had chapels by 1203, but it was long before they established their parochial independence. As late as 1585 there was question whether the rectories of Grasmere and Windermere should pay some acknowledgment to the vicarage of Kendal. At Hawkshead there was probably only a handful of inhabitants till the area came into the hands of Furness Abbey, who established a grange there. Though a chapel was founded early, it became independent only in 1578. The little chapel of Cartmel Fell, one of the most interesting and attractive in our area, was built about 1500 on a fellside looking towards Witherslack, probably mostly with money from expanding wool trade. It became parochial even later. Architecturally most of the churches are not impressive, largely because of the poverty of the area in early days; but Cartmel Fell contains attractive medieval glass (probably originally in Cartmel Priory) and other ancient fittings.

An important development in local life was the spread of monasticism in the course of the twelfth and thirteenth centuries, though most Lakeland monasteries, like Furness Abbey, Cartmel Priory, St. Bees Priory and Holme Cultram Abbey, lay outside the present National Park. Calder Abbey, still well worth a visit, belonged to the Cistercian Order and, like Furness, occupied a sheltered and lonely valley. Its remoteness and moderate means gave it an uneventful history. Of Westmorland's only abbey, the Premonstratensian house at Shap, there are interesting remains, notably a tower copying that at Fountains Abbey in a small way, and recently well restored.

In comparison with the monasteries of more prosperous regions of medieval England, those of our region were neither numerous nor, with the exception of Furness Abbey, influential in national affairs. But they certainly helped to develop the countryside. Remarkable is their contribution to economic development. Local monastic houses exploited the good-quality iron ores of western

Cumberland and Furness (on the borders of the existing Park). More directly important from the thirteenth century onwards was the steady development of the wool trade which led to utilisation of great areas of the present National Park as sheep-walks. Fountains Abbey, the richest Cistercian house in England, built up considerable estates here for this purpose, notably around Keswick and Borrowdale; and Furness had estates around Hawkshead and Coniston. An unusual side of monastic economy here is presented by the saltpans, maintained along the coast to supply the salt, often so difficult to obtain at this time, and so valuable for the salted meat and fish which was a staple diet for much of the year in an age that knew not refrigerators. In bad seasons the local supply of corn often failed and we find Furness and Cartmel importing it from their possessions in Ireland. The monasteries helped and developed social life among their tenants in a variety of ways, providing some educational facilities, relieving poverty and illness, and guarding against floods.

By the end of the fifteenth century many people of the Park area were appreciably richer than they had been in early medieval times, thanks in no small measure to that unlovely thing the Lakeland sheep. By then, writes Professor Carus Wilson:

'Westmorland was famous for its cheap but durable "Kendals" made from the wool of the hardy sheep that roamed the Lakeland fells . . . . The industry spread far beyond Kendal and the valley of the Kent, penetrating the furthest recesses of the Rothay and the Brathay where there was an abundance of water-power and also of cheap land, barren and rocky, worthless for agriculture but suitable for the erection of tenters for stretching and drying the cloth. In the parish of Grasmere alone, in the place of the one manorial fulling mill of the early fourteenth century, there were now at least twenty such mills, some in spots as remote as on the becks descending from Stickle and Easedale tarns . . . vigorous Lakeland carriers ventured south each winter with cloth for the Italian galleys, taking home with them winter luxuries for the Westmorland housewife—oranges, nuts, wines and dried fruits—in addition to the raw materials of industry.'

The sixteenth century brought the major religious changes of the Reformation but little else that was new. For some reason local buildings and activities were minimal. Border warfare still flared up. But in 1707 the union of England and Scotland brought lasting relief. Also, for reasons not yet clarified, in the latter half of the seventeenth century Lakeland agriculture enjoyed a boom unparalleled until our own day. Many farmhouses great and small were rebuilt with the massive roofs and chimney-stacks that still survive and furnished with sturdy carved oak chests, settles, court cupboards and spice-cupboards, not to mention humble but attractive Bible-boxes. Of the massive tide of tourism which today poses such problems for the Park, there was as yet not the slightest sign.

7
Social and Economic Change

by G. P. JONES

SOCIAL and economic change in the Lake District itself has been delayed by various factors, of which one was nearness to the chronically disturbed Border. Another retarding factor, though it would be possible to exaggerate its importance, was the remoteness of the district, in an age of poor, slow and costly transport, from large centres of population and important markets. Down to fairly recent times, as the least imaginative visitor to Kentmere, Cockley Beck or Crummock may realise, life in the region was marked by isolation and by a relative simplicity of manners.

The lie of the land and the climate in times past inevitably determined the nature of farming and industry in the Park. Little wheat could be grown; barley was commoner; but the main food corn was oats. The oatmeal was consumed in the form of flat cakes called clap bread, or of porridge which, with milk, distinguished the diet of the northern labourer from that of the bread-and-cheese-eating farm workers of the southern counties. Potatoes, grown in the south just outside the Park, near Ulverston, as early as 1673, spread during the next century as a field crop in the valleys and on the lower slopes. There, too, were pastures for cattle and sheep, but the mainstay of the flocks was the vast extent of ground for grazing on the higher fells. For centuries society in the district was fundamentally pastoral, and though cattle (at first of the long-horned kind) were important, there can be little doubt that sheep were more so, especially as the foundation of the main textile industry.

The ancient breed, now surviving only in traits of sheep improved by cross-breeding (Plate VI), was probably something like the Black-faced heath sheep once widespread in northern Europe or the Herdwicks, which, according to a very dubious tradition, were of foreign origin. Certainly the fell sheep, able to thrive where other breeds would have died, were small and their wool was poor. About 1350 it was valued at only five marks the sack, compared with ten and a half marks for Shropshire wool and twelve marks for that of Herefordshire. It was nevertheless worth exporting; and the Abbot of Furness was caught in 1423 sending it direct to Flanders

Social and Economic Change

contrary to regulation. It was also good enough for home manufacture though not of the best cloth. With the diminution of forests and, at some date unknown, the disappearance of the wolf, and with the expansion of the domestic cloth industry, it may be presumed that the numbers of sheep, despite heavy mortality at times, greatly increased. Monastic houses had large flocks: Furness Abbey, for example, had sheep-walks as far away as Upper Eskdale and Borrowdale. Great nobles, too, were sheep owners: Lord Dacre had over 5,000 in 1534. The flocks kept by ordinary sheep farmers were, of course, much smaller, but in the early nineteenth century some men could be found who owned or rented hundreds of sheep and a few had 2,000 or more. With the introduction of turnips and other foodstuffs the aggregate number of sheep within the National Park probably increased greatly. In the large Cumberland parish of Crosthwaite there are said to have been 30,000 towards the end of the eighteenth century. In Westmorland alone, in 1932, there were over 440,000.

The original lay-out of the land on which the inhabitants grew their crops and pastured their stock resembled that which long prevailed in other parts of northern England and in Scotland. Two features of it were the holding of land in dales, or strips, in open fields, and a distinction between infield and outfield, the former being the land nearest to the homestead, whose soil received all or most of the scant manure available, while the latter, further away, was left to unassisted nature. Part of it was sown with corn year after year, until, exhausted, it was allowed to revert to its primitive condition of rough pasture. The open fields and dales have been shown by recent researches of G. Elliot to have existed, at one time or another, in at least 220 out of 288 Cumberland townships. They can be traced in Furness, for example in Broughton, Subberthwaite and Coniston, and in Westmorland, for example in Mardale and Wet Sleddale. They could still be seen in Torpenhow, just outside the National Park, in the early nineteenth century; but they had been slowly disappearing for a long time. Thus in 1578, the holdings of Percy tenants in Wasdale Head consisted in each place of a little garth adjoining their dwellings, and dales, from 3 to 10 acres in extent, in the open Wasdale Head Field. In Nether Wasdale, on the other hand, only six out of 46 tenements included land in Wasdale Field, and in Eskdale only two out of 36 tenements included land in Eskdale Field. That is, the open fields were being cut up into closes of arable and meadow, and as that happened the pieces were enclosed either with hedges or with drystone walls (dykes), so that the countryside was beginning to show the pattern with which we are familiar. The next great alteration, with stone walls running for miles along or across the contours of the fells, came with the enclosure of the common pastures in the eighteenth and nineteenth centuries.

Cumberland, Westmorland, and Lancashire north of the Sands were never organised on a manorial basis to the same extent or in the same way as the Midlands and the south and, though serfdom was not unknown in these northern parts during the Middle Ages, it is likely that the proportion of tenants who may be regarded as free was higher than in the country south of the Trent. An undetermined, but large, proportion of the inhabitants in Elizabethan times held the land by a customary tenure of obscure origin commonly known as tenant-right. It may well have evolved not from villeinage, like the copyhold tenure further south, but from the more honourable status of drengage. It varied in detail from place to place but was everywhere broadly similar. The Borrowdale custom required tenants to be ready, at the order of the Warden of the West March, to serve in arms against the Scots; their land was secured to them and their heirs while they paid the customary rent and services; and they were subject to a payment on the death of the lord or on change of tenancy. On the ground that the union of the English and Scottish crowns ended the need for Border service, James I tried to abolish tenant-right, but the 'statesmen' succeeded, though at great cost, in defending their tenure. In an age of steeply-rising prices, such as the sixteenth century was, landlords dissatisfied with fixed customary rents tried to exact arbitrary fines on change of tenancy and to compel the yeomen to accept leases instead of their 'estates of inheritance'. In some instances they had their way, but the statesmen survived as a class through the seventeenth and eighteenth centuries, some prospering and extending their holdings and buying their freedom from services and succession payments, while others went to the wall. Wordsworth believed that by his time the number of statesmen had greatly declined while the size of the holdings had greatly increased; and certainly by 1829 the statesmen formed a minority of the occupants of land. In some places, such as Borrowdale, their proportion was indeed over 60 per cent, but in the parish of St. John's Castlerigg and Wythburn it was under 25 per cent. In Furness and Cartmel as a whole it was 31·7 per cent; in Westmorland 35·4 per cent; and in Cumberland 37·8 per cent. The decline continued in the nineteenth century: it is believed that there were 899 statesmen in Westmorland in 1829 and only 439 by 1885.

The statesman's holdings commonly ranged from 40 to 100 acres, though there were a few larger and many smaller than that. They were usually worked by the statesman and his family; and if there

Social and Economic Change

were too many sons for the purpose some became workers on other men's farms and hoped in time to have farms of their own. There was little or no social difference between the Lakeland farmer, whether statesman or leaseholder, and such workers as he employed:

> And o' fare't alike—beath maister and man
> In eatin' and drinkin' or wark;
> They turn'd out at morn and togidder began
> And left off togidder at dark.

That was one reason perhaps why the region escaped the conflicts and rick-burning which troubled the south about 1830. It may be noted too that, hard and simple as rural life may have been in these parts, the inhabitants were by no means sunk in ignorance: in the early nineteenth century, according to Lord Brougham, one Westmorland child in seven attended school, as compared with one in 24 in the Midland counties; and Dr. J. D. Marshall has shown that, between 1750 and 1830, five out of every six men and two out of every three women married in Lowick and Blawith could write their own names.

In the main, the industries of this corner of north-western England were and are on the edges of the Lake District. Nevertheless, the district itself had, in addition to its wool, some mineral wealth, especially copper; it had very good slate, *e.g.* near Coniston, in Langdale and at Honister, and, in the region of Kendal and Shap, granite; and it had, in an age when several industries were in bondage to forest and stream, advantages in its woods and coppices and its innumerable becks, used to drive hundreds of mills for grinding corn, fulling cloth, working hammers, making paper or turning bobbins. The exploitation of copper began in 1564: then and in the following years miners and metallurgists were brought from Styria by the Company of Mines Royal, a large part of the capital being subscribed by an Augsburg firm, which however withdrew in 1577. Though the undertaking was continuously hampered by financial difficulties, considerable quantities of copper were mined, in Newlands, Borrowdale, Caldbeck, Grasmere and Coniston, and smelted near Keswick. For a time boom conditions prevailed, involving hundreds of men, miners, furnacemen, labourers, lumbermen, hauliers and others; but, though mining was active at times thereafter, the boom was over by 1650. There is room to believe that the population of Keswick was doubled between 1560 and 1590 and had by 1688 fallen to the 1560 level. In the same region, in the eighteenth century and later, there was spasmodic activity in the mining of plumbago, or 'wadd', useful in several ways in industry, as for example in reducing mechanical friction, and good material for lead pencils, a manufacture still carried on in Keswick.

Ancient iron-making has left traces still visible in slag heaps on the western shore of Coniston Water and many other places in Furness. During the seventeenth century forges came into operation at Low Wood, Cunsey, Ulpha, Colwith, Coniston, Backbarrow and elsewhere, and during the eighteenth relatively large-scale undertakings were busy on a variety of sites both within and beyond what is now the Park boundary, including Newland near Ulverston, Penny Bridge, Leighton near Arnside, Duddon Bridge, Backbarrow (3 miles below the foot of Windermere) and Lindale associated with the great ironmaster John Wilkinson, who however sought wider scope in Denbighshire and Shropshire about 1750. The iron was produced by means of charcoal, the burning of which was a common occupation in the Furness woodlands. With the spread of superior techniques based on the use of coal as fuel (though Backbarrow was producing charcoal iron in fairly recent times) the Furness industry, very important in its day, came to an end. (See also page 73).

There were a few cotton mills in the district in the later eighteenth and early nineteenth centuries, for example at Backbarrow and Spark Bridge 3 miles to the west, but the major textile industry depended on wool, the yarn being spun in practically every farmhouse and cottage and available for one or more weavers in nearly every parish. There may still be seen in some farmhouses, for example at Hodge Hill in Cartmel Fell and Pool Bank on the other side of Winster below Cowmire Hall, galleries in which, when the weather allowed, the spinning was done. The main centre for the finishing processes and for marketing was Kendal, 25 of whose cloth merchants were taking or sending their wares as far afield as Southampton in 1553. During the seventeenth and eighteenth centuries the trade in Kendal 'cottons', which were woollen fabrics, though subject to fluctuations was active, the wares being exported to Africa and the American Colonies; but changes in fashion, and especially the expansion of the West Riding manufactures, killed it in the nineteenth century. Similarly the manufacture of stockings and other knitted goods, widespread to the east of the National Park in Ravenstonedale, Orton, Kirkby Stephen, and the Kendal region, perhaps at its peak about 1750 and still brisk in 1800, had greatly decayed by 1830.

Among the more notable changes in the life of the district during the past two centuries there may be counted, first, a growth of population. Westmorland, which may have had about 27,500 inhabitants in 1700, had 40,800 in 1801 and 65,400 in 1931. During

Social and Economic Change 69

that time the population became a little more urban, as is indicated by the fact that in 1801 about 17 per cent, and in 1931 about 22 per cent, of the county population lived in Kendal. The increase, however, was not general: Askham, Bassenthwaite, Orton and Ravenstonedale are instances of parishes which were more populous, the last two a good deal more so, in 1801 than they were 150 years later. There can be little doubt that not only the nearer industrial and urban centres, such as Workington, Carlisle and later Barrow-in-Furness, but Manchester and even London had during the eighteenth and nineteenth centuries, as earlier, drawn from the Lake District young men and women seeking better opportunities than their native places afforded.

A second change was land enclosure. Comparatively little was effected in the eighteenth century but a good deal in the period of the Napoleonic Wars. In Westmorland, for example, out of a total of some 122,000 acres, about 13,260 acres was enclosed between 1765 and 1799, and about 44,500 between 1800 and 1819. There still remained nearly 32,000, or rather more than a quarter, to be enclosed after 1850. There is no doubt that here as elsewhere enclosure was an aid to progress, but experienced observers, such as Thomas Wilkinson of Yanwath and William Pearson of Borderside, pointed out that in some instances the expected gains did not balance the losses, especially to smallholders and cottagers; and it should be remembered that many of these were also losing opportunities to eke out a living as factory-produced yarns, cloths and hosiery were superseding home-produced goods.

Certainly there was an improvement in farming as yeomen and leaseholders slowly adopted the new methods introduced by reforming landlords such as Dr Robert Graham of Netherby and J. C. Curwen of Workington. One improvement, introduced by Philip Howard of Corby about 1755, was the use of turnips as a field crop, which enabled cattle to be kept alive through the winter and to be fattened for market, and also obviated the autumn slaughter of sheep. Meanwhile another activity of some landlords, to be carried further as a matter of public policy in our own day, was altering the scenery; that was the planting of trees by the scores of thousands. One eminent planter was Dr Watson, absentee Bishop of Llandaff, who lived at Calgarth, and who is reported to have said that should his laurels fade he might be remembered by his larches.

Contemporary with these changes were developments in transport and communication, important for three reasons. First, they helped to complete the process which the defeat of the rebel northern earls in 1569–70 had greatly advanced, of integrating the district fully in the realm; secondly, they were essential for its economic progress; and thirdly, they opened the district to a flood of tourists and visitors. Parts at least of the Roman road system, to which monastic houses had probably added, sufficed for cattle drovers and others in the Middle Ages. In Elizabethan and Stuart times the justices of the peace had done something to compel reluctant parishes to maintain highways subject to an increasing strain from carts and packhorses. The latter were the chief means of conveying goods: rather more than 354 horse-loads normally were moved into and out of Kendal every week. Enclosure brought about some improvement of local roads, but the chief means was a series of Turnpike Acts after 1750. The first turnpike trust, in 1752, had charge of the road from Kendal through Kirkby Lonsdale to Yorkshire; the second, in 1753, of the road through Kendal and over Shap to Eamont Bridge. In 1761 a trust was set up for the road from Kendal through Ambleside and over Dunmail Raise to Keswick and thence over Whinlatter to Cockermouth. Two years later there was a turnpike, very steep in parts, from Kendal through Newby Bridge, Bouth and Penny Bridge to Ulverston, but a better alternative for those coming from the south and wishing to avoid the Leven Sands was provided in 1818 by a road from Levens to Greenodd. Thus the way was open for carriage traffic. The first stage coach ran over Shap in 1763, not very quickly; the North Mail in 1793 took six hours for the journey from Kendal to Carlisle. An addition, important in its day, was the canal, now in its northernmost stretch empty and grass-grown, which reached Kendal in 1819, enabling the coal of south Lancashire to be brought in (and the town soon to be gas-lit), and the lime and other goods of south Westmorland to be carried out. But the real revolution came with the railways. The route from London through Lancaster was open to Carlisle by 1846; the Furness Railway (1844–57) reached Coniston in 1859; Wordsworth was protesting violently against a line from Kendal to Windermere in 1844; the line from Cockermouth to Keswick and Penrith was carrying passengers along the shores of Bassenthwaite Lake by 1865; and the line from Settle through Appleby to Carlisle was open in 1876.

Some visitors to the Lake District in the eighteenth and early nineteenth centuries were seeking, in country air and mineral springs, such as Witherslack 'Spa' and Shap Wells, remedies for real or fancied ailments. Many, especially after Thomas Gray's tour in 1769 (see page 122), were in search of picturesque or romantic scenery. Not a few came because it was the fashion to travel, and the Lake District was more accessible than Switzerland. When the railways provided for the many opportunities hitherto available only to the few there can be little doubt that one cause of the influx was a desire to escape, if only for a time, from the noise and smoke of industrial towns. The peace and beauty of the country induced many to become permanent residents, so that early in the nineteenth century villas were already multiplying in the neighbourhood of Windermere and Derwent Water. To supply their needs, and also to cater for holiday-makers, an increasing number of innkeepers, domestic servants, shopkeepers and tradesmen gathered in expanding centres such as Bowness, Windermere and Keswick. The population of the last rose from 1,350 in 1801 to about 4,500 today. Thus a new element, in the form of what may be called a tourist industry, altered the pattern of life in the district. The old domestic textile industries have practically disappeared; the forge hammers have long been idle; charcoal-burning is no longer common; gunpowder is not made now in Langdale or Black Beck. Mining and quarrying have been, though spasmodically, active; they employed about 7,000 men in Cumberland and Westmorland at the opening of the present century. Industries little known before, such as paper-making, have extended and flourished. Milk, now largely that of Friesian cows, is carried daily to the creameries in large quantities, and few farmers' wives now use churn or cheese press. The sheep remain, and, despite highway improvements, motor cars and electric light, there still survives something of the old character of a pastoral society.

8
Industrial Archaeology

by J. D. MARSHALL *and* M. DAVIES-SHIEL

ONE of the earliest investigators to use the word 'archaeology' in connexion with industry was a versatile Cumbrian industrialist named Isaac Fletcher, well known about a century ago for his work in railway promotion and iron-making. In 1876 he published a study entitled *The Archaeology of the West Cumberland Coal Trade.* His subject, of course, lay outside the area of the present National Park, and little did its author dream that local historians and technologists would one day reveal innumerable sites and traces of industry in the remotest fells and dales. The comparative lateness of the revelation was not of course Fletcher's fault, and until recently local historians have tended to neglect those sites and traces which are 'modern' in terms of a wide time-scale.

The National Park area has had a variety of resources which have aided the growth and maintenance of local industries: minerals, wool, water-power, and wood-fuel. Their remaining traces take many forms also: mounds of cindery slag in dales and by lakes, representing medieval or early modern iron-smelting by the crude bloomery process; deep and tortuous workings made in and after the Tudor period in pursuit of copper or lead; ruinous or converted stone buildings which were once bobbin mills; and—awe-inspiring if sometimes unlovely—the remains of slate or granite quarries at Tilberthwaite or Threlkeld.

Some, at least, of these early industries arose to meet a largely local or restricted demand, and this was certainly true of the iron bloomeries. Furness Abbey is known to have been interested in the smelting of the local haematite iron ore, which was carried into the neighbourhood of Coniston Water along the road which runs by above Ulverston and into the Crake Valley. In this area was the vital charcoal fuel, obtained from the woodlands which clothed the sides of these southern dales. The monks, and doubtless others, evidently learned to conserve these woods, and the monastic or late Latin word *colpecia*, 'wood for cutting', giving us the word *coppice*, is indicative of their preoccupations. As was observed in 1537:

'There ys moche wood growing in Furneysfells in the mounteynes ... where in the Abbotts of the same late Monastery have been accustomed to have a Smythey and sometyme two or thre kepte for making of Yron to thuse of their Monastery ...'

For three centuries the woods were cut and conserved systematically, the resulting cordwood supplying the charcoal-burners, and also basket-makers, tanners, wood-turners, swill basket-makers, saddletree-makers and other trades. Nor was the scene a sterile wilderness as a result; the 'Byrk Holey Ashe Ellers' and 'lytell short okes' which are typical coppice trees in this area together produce a rich and subtle pattern of colouration in spring and autumn, to create the delicate yet characteristic scenery of the Furness Fells not far from Newby Bridge, at the southern extremity of the Park. This corner of Lakeland is largely man-made, and it fostered industry after industry. So much for stereotyped notions around the latter word.

The bloomeries were primitive hearths, perhaps diminutive temporary furnaces with a draught made by grinding labour at hand-bellows, using the local charcoal. Iron was crudely reduced or smelted at these spots for generation after generation, the work producing heaps of slag several feet high and many yards in length. There is a very fine example at Springs, in the caravan park south of Coniston Old Hall, and another, better hidden (and less liable to ignorant destruction), just south of Ruskin's Brantwood on the other side of the lake. These sites are quite well known. Less well known are up to three hundred others, scattered throughout the fell country, but most frequently from Windermere to Wast Water. There are numbers in Eskdale, perhaps because this was within packhorsing or carting distance of the Cleator iron ore mines. Metallurgists have been investigating the work sites and analysing the slags, and so our information will become more precise. Amateurs should not dig into the heaps, for the digging destroys vital clues.

This primitive industry gave way, in the course of the seventeenth and eighteenth centuries, to much more sophisticated iron-making, employing water-powered blast equipment and substantial stone-built furnaces with an associated complex of charcoal sheds and other buildings. There is a very fine site, dating from this period (1736), at Duddon Bridge—in fact this is really the 'Coalbrookdale' of the north-west as far as appearances go, although the Duddon furnace is heavily overgrown and has been in serious danger of collapse, the result of shameful neglect. There is another interesting but less impressive site at Nibthwaite near the foot of Coniston Water. Here iron was forged as well as smelted from the local ore;

the greater furnaces like Duddon sent their pig-iron several miles to local charcoal forges to be reheated and hammered, the resulting decarburisation producing bar or malleable iron, which was shipped away by coastal vessel to the south-west. The forge slag is heavier than the bloomery variety, and is rich in iron; there are plenty of examples in the roadside walls near Nibthwaite. The lumps are known as 'mossers', and the specimens here date from the mid-eighteenth century. Unfortunately the small local forges at Force near Satterthwaite, Coniston, Cunsey, Spark Bridge and Backbarrow have left comparatively unimpressive traces, and by an irony, one of the economically less important forge sites, that at Stony Hazel near Rusland church, has been much more completely preserved. This has now been leased to the University of Lancaster for careful examination; it was first used in 1719.

This local industry became a small and temporary part of the national economy in the eighteenth century. What of Lakeland's textile industries of wool and linen? The rough, rather low-grade Kendal 'cottons' (in reality woollens) were fairly well known by Shakespeare's time, and packhorse-driving distributors known as *Kendalmen* were familiar to southerners. Most of the traces of this industry are documentary and topographical rather than archaeological. The well-known 'spinning' galleries of Lakeland, more correctly termed slate galleries, were ideal for the purpose of drying washed fleeces. They preponderate in the areas of damper air in southern Lakeland.

However, a newly recognised class of kiln, now known as a potash pit, has fairly recently been brought to light, through the work of M. Davies-Shiel and W. Norris. These kilns form an integral part of the pattern of the early woollen industry to be found throughout Lakeland, often where there is a Tudor settlement, with its fields as tiny enclaves in the wilder setting of the lowland fell country. The older kilns burnt timber, but later kilns were constructed to burn green bracken. The potash thus produced was mixed with burnt lime, water and tallow, to produce *lyes*, the essential soapy ingredient used in all early fulling mills of northern Britain. The kilns occur in dense clusters in each valley, and where valleys naturally gather to a central point, one may search for a woollen centre. This is clearly true of Kendal and it may well be the ease with which the lyes—and the dyes—were gathered for the fulling mills that caused Kendal to become a centre for the woollen trade here by the fifteenth century. Accordingly, one can stand by the banks of the upper Kent or the Rothay and listen, mentally, to the heavy thump of the stocks, as they once fell on the cloth rolls saturated with lye and water (see p. 68). Fulling mill sites have left little or no trace, unhappily, and most of the recognisable water-power establishments of Lakeland belong to a later age: corn mills, which are legion, woollen mills, paper, bobbin, flax and cotton mills, and, in a few out-of-the-way places, sickle and spade forges. Gunpowder-making plants, at Elterwater, Black Beck near Rusland, and Low Wood near Haverthwaite, all made extensive use of water-power during the last century, as did some of the metalliferous mines of the central Lake District. None of this need surprise us, for the water resources of the region, for this purpose at least, are vast indeed, and the river Derwent, in full spate, has been reckoned to carry as much water as the Thames in normal flow. Such sources of kinetic energy were utilised by Lakeland water-wheels to give well over 30,000 brake horse-power in the mid-nineteenth century, shared between many hundreds of power-users.

Unhappily, many of Lakeland's corn mills have been transformed to other uses, or even demolished. The mill at Boot, in Eskdale, and that at Whitbeck, on the western side of Black Combe, are picturesque and interesting relics.

There are, however, some fine water-wheels still to be seen at Isel, north-east of Cockermouth, at Wythop, at Muncaster, Comb Gill in Borrowdale (Plate VIII), and Witherslack in the south. An overshot wheel is shortly to be rebuilt on the old manorial corn mill building at Ambleside in order to enhance the attractions of the area. On the other hand, there are not so many visible remains of a trade that was at one time more important here than anywhere else in Britain, that of bobbin manufacture.

There are about 50 bobbin mill sites in the Park, representing either smallish units or quite large plant. The size of the initial building depended more on the amount of suitable local timber available than on any other factor. For instance, one of the larger bobbin mills was attached to, and intrinsically part of, one of the earliest large cotton mills in the area—that at Barley Bridge, Staveley, working in 1789. Another bobbin mill, built about 1840, adjacent to another cotton mill site that closed, was the so-called Horrax Mill at Stock Ghyll, Ambleside. This is now partly converted to holiday flatlets, which can be seen as one climbs the Kirkstone road. This establishment manufactured a surprising range of wooden articles from pill-boxes to hatstands, and made enormous demands on the ash, birch, alder and willow trees of the local woodlands. It was at its most productive a century ago. Its major products, the bobbins, were sent all over the United Kingdom. The Stott Park mill, which commenced operation in 1836, drew its water supplies from High Dam, a charming tarn on Finsthwaite Heights which is often regarded as one of the beauties of this part of Lakeland. The lathe shop and the coppice sheds at Stott Park have been seen as typical features of an industry which was once the most important in the Cumbrian countryside.

Bobbin mills were situated in localities as far apart as Howtown on Ullswater, Stainton near Ravenglass, Ulpha in Dunnerdale, and Garnett Bridge, Longsleddale. There is one plant still in use as a bobbin mill, at Spark Bridge, north of Ulverston. At Greenriggs Mill in Underbarrow, west of Kendal, there still remains a fine overshot water-wheel on the outside of a mill that has been at various times a fulling mill, a corn mill, a bobbin mill and a sawmill. It ceased working only recently. The earliest bobbin mills at Staveley appeared in response to the Lancashire demand for cotton bobbins, and one of the largest mills there was eventually taken over by Chadwicks, a Bolton spinning firm.

Hence the industry was a creature of the industrial revolution It. boomed between the 1830s and 1867, when restrictions on boy labour put the mill proprietors at a disadvantage. Fire, Scandinavian competition, scarcity of labour—all these things galled the remaining proprietors. Yet the mills, for all their dreary exploitation of boys and young men, kept dales communities alive and provided a challenge to local resourcefulness. Some of the ingenious machinery used was of local invention, and the woodworking machinery firm of Fell, whose works are still to be seen at Troutbeck Bridge, was established in consequence.

Another important industry, that of gunpowder manufacture, has left locally numerous traces. This employed many dalespeople, and it arose to satisfy the demand of northern quarry owners for supplies of blasting powder. Subsequently, like the bobbin industry, it began to meet the needs of much wider markets. Perhaps the most interesting former works site in the National Park proper is at Elterwater, in an area which is now a holiday estate. This mill was started in 1823–4 by Messrs Huddleston, and it soon made extensive use of water-power to operate the successive preparatory processes of incorporating, pressing and 'glazing' the gunpowder. The various stages of manufacture can still be traced through the site with the aid of a diagram in *The Industrial Archaeology of the Lake Counties*, by the present writers, although visitors should not invade this site or any other without courteous enquiry. The processes had to be kept separate for fairly obvious reasons: an explosion might otherwise spread through the works with devastating results. Isolated explosions were not uncommon in these dales works (there were others, on the edge of the Park, at Low Wood near Haverthwaite and Black Beck near Rusland), but the dangers did not prevent many Lakeland people from 'addling' a living in them. Elterwater works, or its owners, also altered the landscape in an intriguing and little-known fashion, in their extensive flooding of Stickle Tarn, high up on Mill Ghyll above the New Dungeon Ghyll Hotel, to make an improved water supply (1838–9).

The industry, fathered regionally by enterprising Kendal Quaker-ism in the eighteenth century, was of course attracted by the cheap land, cheap labour and copious water, with the availability of charcoal in the coppice woods and the relative cheapness of transport of saltpetre and sulphur. It was still in a fairly flourishing state within living memory, but had died by the Second World War. The workers' houses at Elterwater village and the well-known clock tower at Low Wood still stand as monuments to it.

Lakeland mining, as we have seen, also used water-power. Perhaps the most grimly fascinating spot in the whole of the Park is 'Coppermines Valley' (the valley of the Red Dell Beck) near Coniston village. The state-encouraged Tudor enterprise, the Company of Mines Royal, was engaged in mining for copper here by the closing years of the sixteenth century, after having operated in Newlands, Borrowdale, Caldbeck and Grasmere. It employed German and Austrian miners to use their special skills in pursuing the fickle veins of copper, and some pits and workings of theirs are certainly in this area. Most of the traces in the vicinity, however, are those of mining carried out in the last century. A few hundred yards upstream from the Youth Hostel is an inconspicuous portal, the entrance to the Deep Level driven by the Barratts, Cornish mines speculators, after 1830. This ultimately led to a vast complex of working at different levels beneath the valley-floor, and hundreds of thousands of tons of ore and veinstone passed this way to the crushing and washing plant on the more level ground near the hostel. Higher up the valley still, meanwhile, are the sites of several water-wheels, used for winding or pumping. The various shafts and levels in this locality should be regarded with the greatest circum-spection. They are extremely dangerous.

Coppermines Valley sustained a community of local workers—many hundreds in all—and miners from Alston or Cornwall. Just as fascinating a community has left its visible evidence at Glen-ridding, by Ullswater, where the mighty Greenside lead mine operated until a decade ago. Lead mining commenced here about two hundred years ago, and was employing hundreds in mid-Victorian times. The rows of workers' houses at Glenridding, dating from the 1850s, were occupied by lead-smelters and miners, many of them from Coniston and Alston Moor also. The mine itself used much water-power, and the patterns of leats and sluices can be partly discerned. Both Top Dam and Red Tarn, a natural tarn on the slopes of Helvellyn, provided supplies, and were enlarged about 1860. The traces lower down the valley are not beautiful. The original workings (pre-1830) are on the fell about three-quarters of a mile up Swart Beck.

These mining areas, in the depths of rural Lakeland, were those most heavily transformed and marked by industry. The lead mines at Greenside and Brandelhow (Derwent Water) polluted the local lakes, but all were well established before the conservation movement found its voice in mid-Victorian times. The quarries, too, have not always enhanced the beauty of the region, but it is a reassuring thought that all but the harshest scars and faces become somewhat softened by time. The vastness of the slate quarries at Tilberthwaite can still impress the visitor, and the underground galleries or levels at Honister can be awesome too. The development of quarrying in fact owes much to the railways which began to penetrate the Lake District in the middle of the nineteenth century, and it was an attempt to throw out a line of track to Honister via Derwent Water's western shore that unleashed the righteous fury of conservationists in 1883. Today railways are regarded with nostalgia, and the National Park faces an even greater threat. Meanwhile, we should remember that Lakeland's industries helped to support a dales population which had to learn the art of survival in singularly inhospitable territory. Without men the Lake District would have looked very different; and it is men that have given it much of its character.

How to Enjoy the National Park

and any assessment of the possibilities of the area must take proper account of the less energetic pursuits, as well as acknowledging the more arduous pleasures. It may be convenient, I feel, to divide the 'Things to do' into three categories—the leisurely or contemplative, the spectator pastimes, and the opportunities for the energetic or daring—but before this is attempted I must say a few words about the most important matter of the weather. (It is considered more fully as a subject in Chapter 14.)

And here a popular fallacy must be quickly corrected. For it is not always raining in the Lake District. In fact, rain is little more prevalent in the fell country than in most parts of Britain, although admittedly when it does rain it can rain very heavily indeed. But the area can be much sunnier than some other vaunted holiday resorts, and frequently when much of the rest of England is covered in rain or fog the Lake District is basking under blue skies.

There are perhaps two main reasons for the completely mistaken impression that the Lake District is more often wet and cold than dry and sunny. One is that the tiny hamlet of Seathwaite at the head of Borrowdale has the unenviable notoriety of being the wettest place in the country; the other is the prominence given on the radio and in the Press to reports of snowbound roads or floods in Lakeland. Undoubtedly the rainfall records at Seathwaite are extraordinarily high but this is only because the hamlet lies immediately beneath the highest land in England, and some of the gauges are sited some distance up the fells where naturally the rainfall is very heavy. But it does not rain much more often in Seathwaite than in other parts of the north-west. On the other hand, there are places in the Lake District where the rainfall is remarkably low. The fertile Eden Valley beyond the north-east boundary of the Park, for instance, is a relatively dry place while Eskdale, one of the western dales, is often quoted as a valley with an enviable record for good weather, even though it lies close to the highest mountains. Indeed, the fact that the Park lies so close to the west coast with many of its valleys opening out towards the sea means that the weather is apt to change quickly, and a rainy morning is often followed by a gloriously sunny afternoon. And even a wet day has its compensations for those who care to seek them out. The fells take on a new grandeur in mist, rain or storm, and the sight of the clouds sailing low across the hills, the new thunder of waterfalls and the clamour of torrents in swirling spate can be strikingly rewarding. And when the rain is over, finished perhaps as quickly as it began, and the first sunlight smiles down from a triangle of blue sky, there is a new magic in the fresh smell of the earth and the woods, the sharpened colours of the fellside and the rocks, and the sparkle of the moisture on crag and gully. These changes that contribute so effectively to the variety of the scenery often provide the main delights of a day in the hills. I have seen the Isle of Man and even Ireland from the summit of Sca Fell, sharp as the hills across the Solway Firth, after a morning of thick mist and heavy rain, and a hundred times have climbed on foot or ski through the clouds and the drizzle into a wonder-world of sunlit snow and clear blue skies. So that even with the weather at its worst, the sun may be only just around the corner.

One is often asked to name the best month of the year for visitors, but this is not an easy question to answer. There is, in fact, much to be said for any month of the year, not excepting November which can be a rewarding month with the reds, oranges and browns of autumn still with us, the crowds departed, the traffic stilled, the views often long and clear, cheery log fires and capacious meals, the thrill of a morning hunt with one of the mountain packs and, with just a little luck, dry bracing weather before the winter snows begin. The winter months can provide a wonderful season, especially for the energetic, and in some years the skiing at times can be nearly as good as in the Alps (although deficient in mechanical transport), the winter mountaineering superlative, and the skating continuous for weeks. One Christmas was memorable. I remember spending Christmas Eve, Christmas Day and Boxing Day skating in the sunshine on the ice of Tarn Hows. In fact, we ate our Christmas dinner—chicken sandwiches washed down with a bottle of wine—with our skates on, sitting on the edge of the ice in a little sheltered bay. The day was windless—at any rate in our little bay—the sun so warm that we basked pleasantly comfortable in our sweaters without coats or windproofs, and the ice was hard, black and almost unmarked. It would be impossible to visualise better skating conditions anywhere in Europe.

And in the spring there are the daffodils, remarkably clear skies, sharp fresh colours, no queues for buses or in the village shops, and the fells largely to yourselves to enjoy, on crisp mornings and lengthening evenings. Autumn brings the most exciting colours of the year—a time when even a bus trip along a main road can thrill with the magical shades of the woods and the fellsides, the golden glory of the dying bracken and the rich red, orange, bronze and yellow and even purple of the hedgerows. And September and October are among the most reliable months of the year, often dry and sunny and rarely ruined by rain, so that this period, as well as May and June at the other end of the summer, is probably as safe a choice as any. The early summer attracts people unaffected by such

How to Enjoy the National Park 83

restricting matters as school holidays and these visitors can often see the still unspoiled charms of the National Park at their best. Easter, the shopkeepers will tell you, brings the active young people for the first open-air holiday of the year, July and August attract the family holidaymakers, September and October the rather more discerning seekers after solitude and less crowded byways, and winter, both early and late, the connoisseurs, anxious to catch something of the real flavour of the Lake District.

It is unfortunate that so many visitors in search of rest, solitude or a quiet contemplative holiday are forced to come back, year after year, in the often disappointing months of July and August. These, quite obviously, are the crowded months when accommodation may be difficult and prices higher, while those who come to escape from and to forget the irritations of city life will not wish to stand in bus queues in Windermere or Keswick. These can also be the least colourful months with the flowers and foliage full-blown or even past their best, months of garish rather than subtle colourings, with road traffic slowed to an irritating crawl and queues not only for the rowing boats and for afternoon tea but also for the more accessible rock-climbs. They can also be the wet months, and many a camper, marooned in a sodden field with curtains of drizzle outside his tent-flap, and lacking the necessary optimistic approach, has thrown in his hand and driven back disappointed to the towns. But this is not to condemn these months altogether, for they are sometimes rewarding and are favoured by more people, although perhaps with no choice, than any other time of year. And if you like the crowds, the bustle of the little towns and villages in the high season, and the fun on the lakes, these are the months to choose, especially, perhaps, if the children are with you. For these are the months of the big sports meetings, the sheepdog trials, the hound trails and the other outdoor events noted below and described more fully in Chapter 12. Grasmere Sports take place in the second half of August. But if you can bear to miss this sort of occasion and have a free choice you are more likely to see the Lake District at her glorious best at other times of the year.

The best age at which to visit the Lake District, it has been stated, is anywhere between 7 and 70 years. But these are unnecessarily narrow limits. I once met a young man of 14 months or so close to the summit of High Crag in the Buttermere fells. He was sitting in a small box, strapped to the back of his perspiring father, and did not seem in the least disconcerted or, I regret to say, even interested in the view. My son climbed his first mountain—Coniston Old Man —at the age of four and did his first rock-climb before he was eight, while my daughter reached the top of Sty Head, with only occasional assistance, at the age of two and a half. While, at the other end of the scale, I have climbed rocks with men of 75 and once met an 82-year-old youngster, sunburned and cheerful, on his own on the top of Helvellyn. And there was a sprightly performer on the ice of Rydal Water early in 1963 who skated in the Lake District for 80 years, while many motoring centenarians have admired the view from Kirkstone summit or Newlands Hause.

I have already suggested the division of the things to do into three categories. In the first, the leisurely or contemplative, I would include the gentle walkers, the indefatigable sightseers, the patient fishermen, and those who must have their coach trip or their game of bowls wherever they happen to be holidaymaking. Obviously, the principal attraction of the National Park is the scenery, and from any one of a dozen or more convenient centres you may see new, exciting or colourful views each day in leisurely excursions either on foot or by coach or car. These walks or drives may be to well-known viewpoints or perhaps to half-a-dozen lakes in turn or to the more dramatic waterfalls or to particularly attractive villages such as Hawkshead. There are quiet walks through the woods, or over the lesser heights, or perhaps along a strip of lake shore, from every centre in the district and, quite apart from enjoying the scenery, the visitor may wish to do the sort of sightseeing at which the Americans excel—seeking out the famous or romantic places, preferably with a camera. There are dozens of these sorts of places, from Wordsworth's homes at Grasmere and Rydal, or his birthplace at Cockermouth, to the Walpole associations at Watendlath, or Beatrix Potter's home at Hill Top, near Sawrey, where she wrote tales about rabbits, mice and ducks. Then there are the tiny Bridge House at Ambleside, the third highest inn in England on top of Kirkstone Pass, the famous Bowder Stone in Borrowdale, the Three Shire Stone, the Ruskin museum at Coniston, many an old church steeped in history, the Roman fort on Hardknott, or perhaps a stone circle or some other romantic mark of man's early settlement of the district. Visitors will be interested in the Abbot Hall Art Gallery and Museum of Lakeland Life at Kendal, and the splendid National Park Centre at Brockhole (Plate IX). And some will want to see the 'Lion and Lamb' rock on Helm Crag above Grasmere, or the floating island on Derwent Water, or the 'Rock of Names' near Thirlmere, or the ruins of Calder Abbey. There's almost no end to the list. In nearly every village there are dramatic waterfalls—Scale Force near Buttermere has a drop of 120 feet—well-known surprise views, and romantic links with the Lake Poets or the famous of long ago, while

How to Enjoy the National Park

in many of the villages there are colourful local customs such as the rushbearing ceremonies at Grasmere and Ambleside and the country dancing in the Langdales. Or you may play your golf, bowls or cricket in or near many of the bigger towns and villages and perhaps within the shadow of the fells, look at exhibitions of pottery, weaving, painting or the local crafts, or try some of the entertainments on a wet day. And you may fish, row, sail or swim in or on many of the larger lakes and picnic on their shores—provided they are not also reservoirs.

So far as 'spectator sports' are concerned there is no lack of choice either, particularly during the crowded summer months, but also at other times of the year. There are, for instance, the many sports meetings where the visitor may watch the traditional Cumberland and Westmorland wrestling and fell racing, the hound trails and the pole leaping—and nowhere better than at the famous Grasmere Sports generally held on the Thursday nearest to 20 August, at Ambleside, and indeed at almost every village in the district. There are also the sheepdog trials—notably at Applethwaite (Windermere), Rydal and Patterdale—the agricultural shows, the horse fairs at Appleby and Brough on the edge of the district, the pony trotting, the swimming races on Windermere, important road and cycle races, the annual mountain trial over the fells, and many, many other occasions when you can sit back and be thrilled or entertained. Or you may spend a quiet morning looking at the glory of the damson blossom in the Lyth Valley or admiring the daffodils where Wordsworth walked.

And now I am coming to the principal recreations of the Lake District, the main pastimes for the energetic or the daring, but these are by no means restricted to walking the fells or climbing the rocks. The fishing in the lakes and rivers for trout, perch, salmon and seatrout can be excellent but there is another type of fishing high up in the fell-backs or in the upland tarns which may require stout lungs as well as patience and more than a little ability. And what greater exercise for the lungs than a day out with one of the mountain packs, following the fox on foot on a crisp autumn morning behind the red-coated huntsman? This, indeed, can be a strenuous, interesting game although nowadays many of the followers are content merely to view the chase from their cars on the road. Bird watching, too, and the study of the mountain flowers or the search for insects or moths can also mean long, steep hours on foot.

But the mountains are mostly for the fell walker, the rock climber, the winter mountaineer and the skier, and on these crags, ridges, snow slopes and icy gullies young men may discover themselves and older people regain their lost youth. The hills are not high compared with other mountain ranges, but they have character, colour and variety and the strongest walkers will find themselves well tested, particularly when low mists or sudden bad weather add spice to the adventure. Set among these hills are bold crags rising up to 700 feet in height where rock climbing of a standard hardly exceeded anywhere in Europe can be practised. Beginners may learn on easy routes and progress to more difficult things, while tuition is given by some of the climbing clubs—and now by the Lake District Planning Board —as well as by a few professional guides. In winter time the north-facing gullies of Great End and elsewhere present problems in snow and ice work, while many of the hills, particularly those in the Helvellyn range, can provide excellent skiing, sometimes for weeks at a time. The climbers have their huts and bothies, and the Lake District Ski Club has its own ski-tow and hut high above Glenridding and welcomes visiting skiers. Skating on the lakes, including occasionally Windermere itself, and nearly every year on dozens of the smaller mountain tarns, completes the winter possibilities—if I add the curling on Derwent Water. Lightweight camping, too, can be practised by the energetic high up in the fells while those preferring a less arduous holiday may camp or caravan in the dales on specially provided secluded sites. This sort of provision is being increased, and chalets and other types of holiday accommodation for families are being made available. There is a chain of excellent Youth Hostels stretching right through the district, and any amount of accommodation of a relatively simple nature.

And now as I conclude this rather brief picture of the possibilities for holiday enjoyment in the National Park I realise how much has been left unwritten. For there is the Royal Windermere Yacht Club, now more than a hundred years old, the Windermere Motor Boat Racing Club and other clubs operating on Bassenthwaite Lake and on Ullswater. (Their addresses, with other local contacts, are noted in Appendix II.) There is excellent water skiing, too, on Windermere and Ullswater—although restrictions on water skiing and high-speed motor boating on Ullswater are in prospect—as well as canoeing and sub-aqua sports.

There are still railway services into the National Park or up to the fringe and excellent bus and minibus transport throughout the district which is well served by motor roads and closely interwoven with quieter lanes leading to hundreds of tracks into and over the fells. And once inside this delectable area, please keep your eyes open for the colour and the beauty and the magic of it all. Study the clouds, get to know the shape and the personality of the moun-

PLATE IX. Brockhole, the National Park Centre, was opened to help visitors enjoy the Lake District

PLATE X. Ancient rights of way make pleasant twentieth-century walking

PLATE XI. True freedom of access is common on most of the Lake District fells

tains, and with the help of other chapters in this book watch out for the wild life, and try to puzzle out the meaning of the old walls, the quarries, the old tracks and the ways of the birds and the sheep. And seek out, in particular, the peaceful, unspoiled Lake District, well away from the popular centres, and talk to the dalesfolk. There's a great deal to learn and it's all important, whether it's the right way to build a drystone wall, or the meaning of the clouds, or an old poacher's tale or the hide-out of a fox. For it all fits in and the whole, with the mountains and the dales and a thousand years of tradition, makes up a heritage which is being safeguarded for you to treasure all the days of your life.

10
Ancient Ways and Modern Walks

by ROLAND WADE

ROADS serve the needs of the day and change with the times. They are repaired, rebuilt, widened, abandoned and replaced; so also are the bridges which taken them over rivers and streams. They are like the old axe which had had five new heads and seven new handles. Scarcely any old Lake District roads retain the form in which they originated. But many can still be traced.

Roman Roads

It seems probable that a Roman road ran from Lancaster, crossing the Kent south of Kendal, to the camp at Ambleside (*Galava*). Here it would join the important road which ran from Brougham near Penrith to the harbour at Ravenglass on the Cumberland coast. This is the most notable of the ancient routes through the Lake District. It climbed steadily along the ridge which separates Ullswater from Haweswater until it reached the summit of High Street (2,718 feet). Then passing close to Thornthwaite Crag—with its tall column, a kind of exclamation mark at the end of the long High Street ridge—it continued down the Troutbeck Valley to make its way to Ambleside. The road then crossed the head of Windermere and went by Clappersgate into Little Langdale. From Fell Foot farm it climbed Wrynose Pass, probably on the line of the present road. On the Duddon side of the pass the route is more accurately known as a result of modern research. It crossed the river higher up than Cockley Beck and made its way by Black Hall Farm—not the way taken by the present Hardknott Pass—to cross the ridge and drop down to the camp and parade ground above the Esk. The camp has recently been restored by the Department of the Environment. The situation is superb, looking down Eskdale to the coast, with the ridge of Sca Fell to the north. But the choice of site was clearly strategic. One wonders if a Roman soldier was ever moved by the beauty of the scene. Or would he only wish he were back in his native land enjoying a beaker of plonk with his pals?

The Roman road continued down the valley, skirting the southern slope of Muncaster Fell, to the camp at Ravenglass, an important port for the Romans. Here are still standing the walls of a Roman bath-house complete with doorways.

Later Roads

Over the years roads were made as needs dictated: roads to the mines which were opened in many of the valleys; roads to the coast; corpse roads for carrying the dead to their burial place in another valley; the road from Dunmail Raise over the fell to Watendlath used by the monks of Furness Abbey when going to their grange in Borrowdale. From early times there was a road along the general line of the present Kendal-Ambleside-Keswick road. A glance at map shows that this is the natural through route. For most of the way to Dunmail Raise it ran a little to the east of the present road, no doubt to avoid the swampy ground in the valley-bottom. Its approach to Keswick was by the Vale of St. John and the straight lane that passes Castlerigg stone circle.

Until recent times almost all the main valleys were cul-de-sacs. The valley road would take wheeled traffic serving the farms and houses, but from the valley-head there would be only mountain tracks and packhorse roads. Kirkstone Pass was an exception; well before this century coaches ran between Patterdale and Ambleside. In more recent times wagonettes from Keswick made a daily round for visitors through Borrowdale over Honister Pass to Buttermere, returning by Newlands to Keswick. The fare in the early part of this century was 5s. for the day trip. Up to the beginning of the last war a few of these wagonettes were still running. The passes were rough unmetalled roads, and between Seatoller and the top of Honister Pass there was a toll road to the slate quarries at an easier gradient than the old road. Now a new well-made motor road attracts a great deal of traffic. It is of interest to note that in the 1902 edition of Baddeley's *English Lake District Guide* he reminds the hirers of private conveyances that 'the Honister Pass is so rough and steep on the Buttermere side as to be practically insurmountable'.

A similar and famous round was known as 'the Round of the Langdales'. 'Char-a-bancs' from Ambleside took visitors daily by Skelwith Bridge, Little Langdale, Fell Foot, Blea Tarn, Dungeon Ghyll, Chapel Stile, Grasmere, back to Ambleside. Baddeley tells us that in 1902 the fare was about 4s.; but for 22s. you could have a private wagonette and pair. The charges included the driver's fee but not the 'feeding of man and horse'. One wonders which was fed the better.

Another road which has been made up in recent times to take motor traffic is the road over Wrynose and Hardknott Passes. As recently as 1932 Abercrombie and Kelly in their Cumbrian Regional Planning Report described the road as only a track and said that, in spite of the bad surface and the frequency of becks crossing the path, Wrynose and Hardknott were occasionally attempted by the rash motorist. There were, in fact, seven water-splashes on the level stretch on the Duddon side of Wrynose. And between Cockley Beck and Broughton-in-Furness there were 22 gates to open—and shut.

These two passes were used during the last war as practice routes for heavy army vehicles. Those did so much damage that major repairs to the passes were essential to make them usable at all. The roads were, in fact, completely remade. The wisdom of doing this was questioned at the time: it is even more in question now because the traffic which the better roads have attracted is causing serious congestion in Little Langdale at busy times and destroying much of the peace and pleasure of those using the passes and the valley-heads both of Little Langdale and of the Duddon and Eskdale. Walks described later in this chapter for these and the other valleys have been chosen so as to avoid road walking as much as possible.

Motor Roads

For many Kendal is the gateway to the National Park and from here runs the main through road to Keswick and beyond to the west coast. Busy with tourist cars and coaches, it is one of the finest scenic routes in England, running along the shores of Windermere with views across the lake to the central fells, through the intimate lake and fell landscape of Rydal Water and Grasmere, and along the full length of Thirlmere. Then quite suddenly the Vale of Keswick comes into view, with Bassenthwaite and Derwent Water framed by the Buttermere and Newlands fells, and Skiddaw showing its full stature.

The other main route into the centre of the National Park is by A66 from Penrith to Keswick, again a busy motor road, now the subject of fierce controversy in view of a scheme to make it into a major through route to west Cumberland. As on the road from Kendal the approach to Keswick brings magnificent views, first into the Vale of St. John and then into Borrowdale and across to the Buttermere fells.

But these are motor roads. They get you close to the heart of the National Park, to the entrance to Langdale and Borrowdale, to the foot of Skiddaw and Saddleback (alias Blencathra—take your choice) and the Helvellyn range; but they are not for walking. From them run the roads which circle the central Lake District, crossing the foot of the main valleys, Coniston, Duddon, Eskdale, Wasdale, Ennerdale, Buttermere and Borrowdale. For these main Lake District valleys are like the spokes of a wheel, radiating from a centre. At this centre the main valley-heads are close enough together for the average fell walker to cross into four or more in a day. The experts will do much more, and for over a hundred years records have been made for long walks over the fell tops. By the early days of this century a pattern for these walks was emerging and Dr Wakefield of Keswick, who later took part in the 1922 Everest Expedition, set up a record for the number of fells ascended in 24 hours and then beat this with a walk which took in twenty of the highest fells. After that he helped another renowned walker, Eustace Thomas, to beat even this record. And right up to the present day others have added even more fells to the list.

These records are possible because the Lake District is compact, only about 30 miles across in any direction. Within this small compass are fifteen lakes, many smaller tarns, and more than fifty fells with heights of more than 2,000 feet above sea-level. The walker does not need to tramp over miles of bog and peat hags to reach the hills, as in some parts of the country. Mostly they rise straight from the valley roads. Nor does he need to be a mountaineer. Some of the best views may be had after less than half-an-hour's walk. Orrest Head with one of the finest views in Lakeland is but twenty minutes' walk from Windermere station.

When walking was serious

In the last century walking was a serious matter. Even a simple valley walk needed thought and preparation. A holiday in the Lake District would normally mean staying in an hotel or lodging-house for a few weeks. In a guide-book published in the latter part of the century advice was given as to dress and baggage:

'What is needed for a lady or gentleman is a fair-sized portmanteau, a Gladstone bag, and perhaps a satchel. The large trunk can be sent on by rail or coach, and the bag can easily be carried on a horse or in a mountain cart. Thick boots are absolutely necessary, and plenty of wraps and waterproofs. The weather is warm in the daytime, but near the lakes quite cool in the evening, so a light shawl should be carried on afternoon excursions.'

What we know as fells were then called mountains, and in the same guide-book an introduction written by Wordsworth for an earlier edition makes comparisons with the Alps. Climbing a Lake

District 'mountain' was undertaken in much the same spirit of adventure as an Alpine peak. 'Ponies', it said, 'can be hired for mountain excursions at about 7s. 6d. per day. Guides for about the same; sometimes 5s. will be found sufficient.' But most of the excursions described in the book were in the valleys; none was complete without a waterfall.

By the early part of this century the 'mountains' had become 'fells'. Fell walking was becoming popular. There were reading parties from Oxford and Cambridge staying at hotels and lodging-houses at the valley-heads, combining rock climbing and fell walking with high intellectual conversation. Seatoller House in Borrowdale was a noted place for these parties. The guide-books were devoting more space to the fell tops and less to the waterfalls. But the 1902 edition of Baddeley, although giving a detailed description of all the main fells, still quotes the prices for ponies and guides. For the ascent of Coniston Old Man, up which thousands now stroll every year with hands in pockets in every kind of dress and footwear, a pony and guide cost 11s. For Scafell Pike the charge was 18s. from Dungeon Ghyll and 10s. from Wasdale Head. 'Ponies may be taken to within half an hour of the summit.'

Valley walks and passes

Windermere. A path for Orrest Head starts nearly opposite Windermere station entrance and winds through woodland until it comes out on to the open fell just below the summit. The lake is spread out like a map: beyond the wooded slopes across the lake are Coniston Fells from Coniston Old Man to Wetherlam; then appearing as if crowded together is the magnificent group of high fells at the head of the main Lake District valleys, Scafell Pikes, Bow Fell, Crinkle Crags and Great Gable, with Langdale Pikes standing out as in so many views from the head of Windermere. Beyond the head of the lake are the fells above and around Ambleside and Grasmere. The walk can be continued over the summit until the path joins a minor road. Turn left to return to Windermere.

Close to the lake on the road from Bowness to Ambleside is another fine viewpoint within a few minutes' walk. Queen Adelaide's Hill is a hillock owned by the National Trust and provides a fine near view of the head of Windermere backed by the high fells.

Claife Heights is the wooded hill across the lake from Bowness. The lake can be crossed by the ferry, and from the west side there are two good walks, one along the lake shore and the other over the fell. For the lakeside walk follow the rough road from Ferry House as far as Belle Grange. Shortly after this at Red Nab the road bends inland but the path continues along the water's edge to Wray Castle. From here it is a short distance to the Hawkshead-Ambleside road and the walk can be continued round the head of the lake to Waterhead. Beyond Wray Castle is the National Trust land of Low Wray; there is hope that a new footpath may be made over this to the head of the lake. From Ferry House over Claife Heights runs a waymarked route created by the Lake District Planning Board. White splashes on rocks, walls, trees and posts show the way. Claife Heights is beautifully wooded and there are fine views of lake and fells.

Great Langdale. For the walker who is bound for Langdale from Ambleside and wishes to avoid the main road there is an alternative by the Rothay Valley and Loughrigg Terrace. Leave Ambleside with St. Mary's Church on the left and cross the meadows to the river. After crossing Miller Bridge turn right and follow the road which runs along the riverside under Loughrigg Fell to Pelter Bridge, close to Rydal village. Do not cross the bridge, but bear left. The road becomes a track when it reaches open land above Rydal Water. Keep along the shore of the lake and then up over a small ridge. This is Loughrigg Terrace, which runs above the south end of Grasmere to join Red Bank, the road from Grasmere to Langdale. The lake with its wooded island has a perfect setting in meadows backed by mature woodlands rising steeply up the western slopes. Beyond the lake is the village and the whole is surrounded by high fells from Fairfield and Dollywaggon Pike to Helvellyn. Part of Skiddaw shows in the dip of Dunmail Raise.

At High Close—a Youth Hostel—the road crosses over into Langdale, one branch dropping down to Elterwater and the other to Chapel Stile for those going to the head of the valley. But there are many variations from Red Bank. One may return to Ambleside by taking the road for Skelwith Bridge and then a track between railings which bears off to the left just before Loughrigg Tarn. This path continues over the southern end of Loughrigg Fell and drops down to Miller Bridge to cross the Rothay again to Ambleside. Loughrigg Fell itself may be crossed by leaving Loughrigg Terrace shortly before it joins Red Bank. Although the height of the fell is only 1,100 feet, fine and varied views can be had by wandering about from one top to another. At the eastern end there is a view down over Rydal Water. Southwards part of Windermere is seen with Esthwaite Water to the right of it and in the distance the flat top of Ingleborough. To the west are Coniston fells from the Old Man round the head of the Langdale valleys to Langdale Pikes. From the top of Loughrigg Fell the path for Ambleside continues southwards to join the track from Loughrigg Tarn to Miller Bridge.

Another fine walk from the top of Red Bank is over Silver How and either down to Grasmere or into Langdale to join the road from Chapel Stile to Dungeon Ghyll. But if you want to see a complete panorama of Lake District fells all round the compass, go on to High Raise, the plateau of which Langdale Pikes are the extremity. It looks nothing, but it looks *at* everything. But a clear day is a 'must' for High Raise.

Coniston. Two walks stand out in popularity above all others, the Old Man and Tarn Hows.

The Old Man is Coniston's very own. As 'The Old Man of Coniston' he appears in maps, but for generations he has been known simply and affectionately as the Old Man; though a visitor might doubt the affection when he sees the way in which his insides have been gouged out and left lying about by generations of mine and quarry owners. To reach the top is a climb only in the sense that the top is more than 2,000 feet higher than the village. The direct way is by a much-used and well-worn track from behind the Sun Hotel, first by the side of Church Beck and then through the quarries. But there is the alternative of taking the Walna Scar road for the first part of the walk. The view from the top is notable for the expanse of sea and estuary, Morecambe Bay and the sands of the Duddon. The full length of Coniston Water is below. To the west beyond the precipices of Dow Crag are the Cumberland coast and the Isle of Man, and northwards a vast sweep of fells from the Scafells, Skiddaw, Helvellyn, round to the Howgill fells above Sedbergh. But to many the Old Man is more than just an 'up and down'; it is part of a fine ridge walk or a place to spend a day or more exploring the caves, quarries and tarns.

For Tarn Hows take the Hawkshead road round the head of the lake: there is a useful footpath adjoining the road. At the start of the steep hill a road goes off to the left. From this road there is a footpath to the right which takes you up away from the road. Tarn Hows is for those looking for the picturesque, an intimate combination of water, woods and fells. There are many viewpoints, with particularly fine views into Langdale. From one point there is a full-length view of Coniston Water. Wetherlam stands out as a giant.

For Coniston itself the best walk is along the east side, where the road follows the shore for much of the way. The Old Man and Dow Crag are prominent across the lake and there are long views up it to the distant fells (Plate III).

The pass from Coniston to the Duddon is by Walna Scar. It starts as a road near the site of the old railway station. Later it becomes a walker's track and makes its way over a dip in the fell at the end of the Coniston range to drop down into the Tarn Beck valley and to the Duddon. On the way down the Scafells, Bow Fell and Crinkle Crags are beautifully framed between Harter Fell and Grey Friar.

Little Langdale. The road from Ambleside to Little Langdale goes by Skelwith Bridge and leaves the Coniston road to drop down to Colwith Bridge, then winds through woodlands until it reaches Little Langdale village. Traffic is now heavy at busy times, but the walker who takes the route from Ambleside to Elterwater suggested in the Great Langdale paragraph can avoid much of it. Leave Elterwater by the Colwith road, which crosses the bridge at the south end of the village. A rough road soon branches off to the right and crosses by Birch Hill into Little Langdale. The tarn lies below, backed by the imposing front of Wetherlam. Higher up the valley the road divides, one branch leading to Blea Tarn and Dungeon Ghyll, part of the famous 'Round of the Langdales'; from the top of the ridge before dropping down into Great Langdale, the Pikes stand up at their most dramatic. The other branch bears off left to Fell Foot farm at the foot of Wrynose Pass. Fell Foot is an old farm and the mound at the west of the house is thought to have been an ancient meeting place. At the head of the pass is the Three Shire Stone, marking the point where Cumberland, Westmorland and Lancashire used to meet. This is a point to start fine fell-top walks: by Red Tarn to Crinkle Crags, Bow Fell and the Scafells; to Coniston fells, Carrs, Brim Fell and the Old Man; to Pike o' Blisco (within the hour) from which there is an extensive view, with the Crinkles and Bow Fell prominent.

The Duddon. From the Three Shire Stone the road drops steeply into Wrynose Bottom and follows the river to Cockley Beck. One road then continues down the valley and the other goes over Hardknott Pass into Eskdale. The walker making for places lower down the valley may avoid the road by crossing the bridge at Cockley Beck and then at once turning left along a farm road to Black Hall. A footpath then follows the west side of the river to Birks Bridge. This little bridge spans the river at a point where it is confined between bastions of rock and below is a beautiful deep pool of the clearest water, where at times sea-trout which have come up from the estuary may be seen resting. In times of flood the water reaches the top of the bridge, and gaps have had to be left in the parapet wall to let the water through and reduce the pressure. Across the bridge is the valley road, but the walker may continue on the west side by taking a track to Birks Farm—no longer a farm since the land was afforested—and then through the forest to Grass Guards. There is then a fine terrace walk between open pastures to Wallabarrow Crag, then a steep cart track down Stoneythwaite Rake to High Wallabarrow, and a minor road to Hall Dunnerdale to join the valley road again. But from Birks Bridge the narrow valley road is equally attractive, with views of the river and into Wallabarrow Gorge, winding and twisting at times between walls and then over open fell until it drops down alongside a series of waterfalls to Seathwaite church. At the head of Wallabarrow Gorge a path crosses the river by stepping stone—Fickle or Fiddle Steps—and works its way up through the forest to Grass Guards and over the fell to Eskdale.

Eskdale. At the top of Hardknott Pass the full length of Eskdale is seen stretching away to the coast with the Isle of Man prominent on a clear day. From here it is only a short walk on to Hardknott Fell. A footpath leads in fifteen to twenty minutes to a boggy plateau round which are ranged a series of tops. As the plateau is reached the Scafell range is seen ahead, with a peep of the Esk Falls where the river tumbles down from under Sca Fell into the lower valley. From the cairn at the western end one looks steeply down to the green meadows of Eskdale with the river winding its way to the sea. From a slightly lower point there is a bird's-eye view of the Roman camp. Beyond Harter Fell, Devoke Water stands high up among the fells. The highest point of Hardknott Fell is on the eastern end and from here the complete range of fells surrounding the head of Eskdale from Crinkle Crags and Bow Fell to the Scafells is seen at its best. The gap of Mickledore between Scafell Pike and Sca Fell stands out sharp against the sky. To the east is the range of Coniston fells from Carrs, the Old Man and Dow Crag to the Dunnerdale fells. There is enough of interest on Hardknott Fell for a whole summer's day—little tarns in folds of the fell, crags, sheltered hollows, and views of coastline and rich pastures, and all the grandeur of the Scafells.

Near the foot of Hardknott Pass on the Eskdale side a track crosses the stream and runs round the base of Harter Fell. This is the track which comes over from Grass Guards in the Duddon Valley. A branch of it gives a fine walk down the south side of the Esk by Penny Hill to Stanley Ghyll. At the foot of the latter is

Dalegarth Hall, which has a lovely example of Lakeland round chimneys. The track continues until it joins the road from Ulpha to Eskdale Green. Across the bridge is the inn known as the King George IV, in earlier times as the King of Prussia (Prussia became unpopular in Eskdale in 1914 and the name was changed). A return can be made to Boot village or to the foot of Hardknott. Alongside the road runs the miniature railway from Boot to Ravenglass known locally as 'Laal Ratty', popular in the summer months. From the road which comes over from Ulpha to Eskdale there is at the highest point a superb panorama of high fells. Harter Fell is close at hand and round the head of Eskdale are Crinkle Crags. Bow Fell, Esk Pike and the Scafells. Beyond are Great Gable, Kirk Fell and Pillar. This is one of the great Lake District views. A short distance from some cross-roads is Devoke Water and on the heathery ridge on the Eskdale side there are fine views down Eskdale to Muncaster Fell and the coast with the cooling towers of Calder Hall standing out prominently.

But Eskdale abounds with splendid walks. Neither Eskdale nor the Duddon has a lake, and this means that there is less tourist traffic than in Langdale and Borrowdale and a quieter time for the walker. Both Boot and the Woolpack are good starting places for walks along the heather foothills on the north side of the valley to Eel Tarn and Stony Tarn. This is the route to Cam Spout at the foot of the Scafells and to Esk Hause. But it is also ideal for a shorter walk, coming down by Cowcove Beck to the river and back by the road to Boot. From Brotherilkeld on the opposite side of the river a path follows the Esk to Throstle Garth at the junction of the Esk with Lincove Beck. The river is brilliantly clear and in it are some of the finest bathing pools in the whole of the Lake District. Where the two streams join there is an old packhorse bridge, recently restored after a serious flood, and beyond it a magnificent waterfall.

Wasdale. Wasdale Head is a noted centre for rock climbers, with Sca Fell, the Napes ridges on Great Gable, and Pillar Rock all at hand. And for fell walkers there are the Scafells, Great Gable, Kirk Fell, and the round of Pillar, Steeple, Red Pike and Yewbarrow. There are also two good lakeside walks. Both should be taken from the foot of the lake towards its head, looking to the high fells and savouring their greatness as you approach them. At Wasdale Hall the road emerges from woodlands on to the open common. Here is Wast Water, the deepest of the lakes, at full length with Great Gable at its head seen as a true gable and not the dome it appears to be from many of the fells. Across the lake are the famous Screes, streaking down straight into the water. The road meanders round promontories and over ridges and there are many opportunities to leave it and wander by the shore. As the road reaches the head of the lake, Yewbarrow shows its steepest flanks. Wasdale Head has a tiny church almost hidden by yews. A famous character named Will Ritson was landlord of the inn more than a hundred years ago, a teller of many tall stories in dialect. So it was said that Wasdale had 't'deepest lake, t'laalest kirk and t'girtest li'er' in all Cumberland.

The other lakeside walk is along the foot of the Screes among the boulders, where the remarkable slope of the scree can be fully appreciated. Near the head of the lake the path joins the track coming by Burnmoor Tarn from Eskdale.

From Wasdale Head, Black Sail Pass climbs up through the gap between Kirk Fell and Pillar to drop down into the head of Ennerdale and then by Scarth Gap into Buttermere.

Ennerdale. Ennerdale is a long narrow valley, the upper part afforested and hemmed in by high fells. The scope for valley walks is reduced, but as with Wasdale there are lakeside walks along both sides of the water, looking to the glorious skyline of Steeple, Scoat Fell and Pillar, with Pillar Rock jutting out lower down the fell. On the south side the path is rough and mostly at the water's edge. At Angling Crag it divides. The high path round the crag is easier than the one which crosses a shelf near the water. At the head of the lake the track crosses the Liza by a footbridge to join the road. Along the north side of the lake is a footpath from the site of the old Anglers Inn to Bowness Hause, about halfway up the lake, and then along the road, which keeps close to the shore.

Buttermere. Buttermere, Crummock Water and Loweswater form a string of lovely small lakes. Buttermere with the ridge of High Stile and High Crag towering above it is a place of great beauty; it remains unspoiled in spite of the traffic over Honister and Newlands Passes and from Cockermouth. There are walks round all three lakes, mostly avoiding the road. From Lanthwaite Hill at the foot of Crummock there is a good view of the whole lake backed by Red Pike and High Stile with Great Gable prominent at the head of the valley. On the way to Buttermere the road rounds a corner at Rannerdale Knotts and the head of the lake comes suddenly into view. Open pastures run down to the shore surrounded by woodlands and in the background are Fleetwith Pike and the high dramatic precipice of Honister Crag. The path on the far side of the lake crosses Scale Beck, which has the highest waterfall in the district.

Borrowdale. Like Windermere, Keswick has an outstanding viewpoint on its doorstep. Castle Head beside the Borrowdale road is only half a mile from the town. From it the whole of Derwent Water is seen with its beautifully wooded islands. At the head of Borrowdale are the Scafells, and behind Keswick the great mass of Skiddaw. The scene is complete and compact, a perfect blending of lake and lakeside woodlands with high fells.

Equally fine is the view from Friar's Crag, a promontory on the shore of the lake reached from Lake Road. The lake at its full breadth fills the foreground. The sharp ridge of Grisedale Pike is a feature, especially when reflected in the surface of the lake on a still day.

A more extended view into Borrowdale is from Latrigg, a spur of Skiddaw immediately to the north of Keswick (Plate I). The lake is the main feature but the background now includes the Buttermere fells and the Helvellyn range. The path for Latrigg goes by Spooney Green Lane, starting from the railway station.

Of the many short walks around Keswick one of the best is to Castlerigg stone circle, not merely for the archaeological interest but for the situation (Plate V). Although standing on level ground of no great height, the monument affords extensive views in which Saddleback is predominant.

Catbells is a mountain in miniature, steep and shapely, rising directly from Derwent Water and a feature in many views of the lake. It is only a simple scramble from the road on the west side of the lake over a smooth slope. It is close enough to the lake to see water and islands and small boats in detail, but high enough to bring the whole massive Skiddaw group into the picture. And to the west is seen the whole of the Vale of Newlands with Hindscarth and Robinson at the head. Catbells and the hamlet of Little Town on its Newlands slope are the setting for Beatrix Potter's *Tale of Mrs Tiggy-Winkle*.

But for the fell walker Catbells is only the first resting place on one of the fine ridge walks of the Lake District, running the full length of Borrowdale over Maiden Moor and Scawdell Fell to Dale Head—though Dale Head can be taken as a separate walk from the top of Honister Pass.

Two of the many valley walks in Borrowdale need special mention: the walk round the lake and the walk to Watendlath.

The walk round the lake starts on Lake Road at Keswick and continues past Friar's Crag on a footpath by the water's edge to the Barrow House boat landing. There is then a short stretch on the road, but after Lodore a path crosses the Derwent by a footbridge and continues past Manesty and Brandelhow on the west side of the lake to Hawse End, the northern tip of Catbells. From here are footpaths mostly through woodlands to Portinscale and so to Keswick.

The Watendlath road bears off left from the main Borrowdale road just before Barrow House. Watendlath is a hanging valley and as the road climbs up the hillside it is in places close to the edge of the cliffs which overlook the Borrowdale road. Here are fine views of the lake and the high fells. The road crosses Ashness Bridge, probably the most photographed bridge in the Lake District, not only for the charm of the foreground with the little arched bridge and wooded surroundings but also for the background provided by Skiddaw, Derwent Water and Bassenthwaite. Watendlath itself is a small hamlet with a tarn. The track goes off to the right over a ridge to drop steeply into Rosthwaite. From the highest point there is a magnificent view up Borrowdale to Scafell Pike, Great End and Great Gable.

Ullswater. On the east side of the lake there is a fine walk between Howtown and Patterdale. If advantage can be taken of the steamer from Glenridding pier to Howtown pier, this is a good way of starting. From Howtown the path goes round the foot of Hallin Fell to the hamlet of Sandwick, then climbs away from the shore, but soon drops again to lake-level and goes round the base of Place Fell to cross Goldrill Beck and join the road back to Patterdale. Throughout there are beautiful views up the lake and to the Helvellyn range of fells. The walk can be combined with the short climb to the summit of Hallin Fell. This has been called a motorists' fell because cars climb the zig-zag road from Howtown and park on the hause, from which it is an easy walk to the top. The lake lies immediately below, and to the south is Martindale, quiet and unspoiled.

Another good walk on the east side of the valley is to Angle Tarn. This starts from Patterdale by the bridge over Goldrill Beck and makes for Boredale Hause, then to the right over the Pikes; the tarn comes suddenly into view. Ahead is the High Street range.

Aira Force is part of the National Trust property, Gowbarrow Park. A footpath starts from the car park near the junction of the Patterdale-Pooley Bridge road with the road to Keswick. It climbs up to Aira Force, a 60-foot-high waterfall, and to High Force where there is a wooden bridge across Aira Beck. You may come down the other side of the beck or the walk can be extended along a terrace to Yew Crag, from which are excellent views.

The main passes for walkers from Patterdale are Grisedale and the Sticks. Grisedale Pass is for Grasmere but is also a useful route for climbing Helvellyn; at Grisedale Tarn a track climbs up on to Dollywaggon Pike and runs along the whole ridge. The Sticks Pass starts at Glenridding and climbs to a height of 2,400 feet over the Helvellyn ridge, then goes down to Thirlmere. Kirkstone Pass carries the main road, and this returns us to Windermere.

One last walk

These are some of the walks in the main valleys; there are many others. And then there are the fells, but fell walking is the subject of Chapter 11. This chapter has dealt with walks which provide views of the fells from the valleys; as a converse one walk may be mentioned in which valleys are seen to advantage from the fells. This is the walk from Honister to Great Gable. It starts behind the quarry buildings at the top of Honister Pass and climbs up to the old drum house and along the slopes of Brandreth. First there is the view down to Buttermere and Crummock with their lush meadows and beautiful woodlands; a little later comes the view down Ennerdale with its bare fellsides and dark blocks of conifers with the shapely lake in the distance; then after crossing Green Gable and scrambling to the summit of Great Gable comes the bird's-eye view of Wast Water with the patchwork of tiny fields around Wasdale Head. And all the time in the other direction are the beauties of Borrowdale, Derwent Water and Skiddaw. If the day is one with cloud about, lifting occasionally for a few seconds to show a shaft of sunlight down one of the valleys, then it will be a day to remember.

11
Fell Walking and Rock Climbing

by J. ROBERT FILES

THE full enjoyment of mountaineering in the Lake District comes from a combination of fell walking, rock climbing and winter climbs on snow and ice, and those who can enjoy all these activities will find the Lake District National Park a miniature mountaineering paradise. Our fells, with their modest elevation, are not mountains on the grandest scale but they have many of the characteristics of mountains; they are much more impressive in appearance than their altitude, measured in feet, would suggest, and, in bad weather, may tax the abilities of experienced mountaineers. But let not this talk of mountaineering discourage anyone who, while loving to walk the fells, has no ambition to climb crags or snow-filled gullies. The Lake District will provide him with enough pleasant and varied walking to fill the opportunities of an ordinary lifetime and those who are more adventurous will find climbing which, at one end of the scale, is little more than a scramble, at the other a challenge to the finest climbers of the age.

Fell Walking

Records of mountain walking appear to date from the latter part of the eighteenth century when a number of visitors to the Lakes were bold enough to ascend Skiddaw and have left impressions of their ascents written in a heroic style which is read today with some amusement. It is more than probable that some of these 'travellers' tales' owe much to what the author thought was expected, in an era of romantic writing in which horror and mountains came to be associated. Many writers of the time, however, show much greater interest in hills as objects to be ascended for the enjoyment of the magnificent views from their summits. Hutchinson (*Excursion to the Lakes, 1774*) climbed Skiddaw and commented that 'the prospect which we gained from the eminence very well rewarded the fatigue'. Other early travellers, among them the oft-quoted Mrs Radcliffe, in 1794, rode to Skiddaw's summit on horseback 'with horses accustomed to the labour', a method of climbing which might conceivably add to the terrors of the journey; and fell ponies were still

PLATE XII. Cumberland wrestling

PLATE XIII. The area has attracted many literary and artistic people—Brantwood was the home of John Ruskin

PLATE XIV. Cumulus cloud developing over High Street. Look out for showers later!

Fell Walking and Rock Climbing

advertised for hire at Grasmere as late as 1914, and were quite commonly used.

Baines, in *A Companion to the Lakes, 1829*, was matter-of-fact in describing the itineraries to Scafell Pike, Helvellyn and Skiddaw, but recommended a guide for the first—a precaution not to be despised in the days when it was exceptional to meet anyone on the hills, routes were not well marked, as we find them today, nor were maps and detailed guide-books available. Green (*Tourist's New Guide to the Lakes, 1819*) accompanied by Otley walked from Keswick to Ambleside over Helvellyn and Fairfield and, on another occasion, over Scafell Pike; and appears, quite rightly, to have been concerned with the pleasures of mountain walking and scenery, rather than with terrors and danger.

Interest in the Lake District became intensified during the early part of the nineteenth century. In 1821 West, in *A Guide to the Lakes*, wrote, somewhat revealingly: 'Since persons of genius, taste and observation began to make the tour of their own country—the spirit has diffused itself among the curious of all ranks'. A change in interest, from an appreciation of the picturesque in mountain scenery to an urge to indulge in fell walking for its own sake, is shown by the publication of Otley's *Concise Description of the English Lakes* in 1825. This is a true guide-book 'directing the tourist through the most eligible paths'. The book contains a section on the Mountains, with routes and details of summit views, a short glossary of names in common use among the mountains, and chapters on the Geology and Meteorology of the district. There is even a list of crags ranging from Pillar in Ennerdale to 'a Raven Crag in almost every vale'. His description of Pillar as 'one of the grandest rocky fronts anywhere to be met with, viewed from above or beneath it has a grand and singular effect' shows a mountaineer's feeling for mountain scenery which will endear him to many present-day climbers.

The popularity of fell walking has increased at an ever-growing rate, partly, as in the early days, as a means of enjoyment for the individual, but largely through the activities of rambling clubs, the facilities provided by the Co-operative Holidays Association and the Holiday Fellowship for fell walking holidays and, during the last 40 years, the establishment of Youth Hostels. Although walking behind a leader, in a large party, may not be one of the higher forms of mountaineering, this method has introduced many thousands to the hills, has taught them something of the joys of mountaineering and has provided the initial inspiration to many to develop their ability for it. There are those who will regret that this invasion of the hills has destroyed one of their charms—an opportunity to enjoy

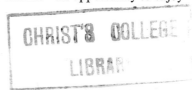

solitude; but the discriminating walker will know where and when to escape the crowds and will not begrudge others the opportunity for a fuller enjoyment of the National Park. Moreover, it might be well to recollect that as long ago as 1894 Haskett Smith wrote of winter:

'There is hardly any time of the year at which a trip to Lakeland is more thoroughly enjoyable. In the first place there is no crowd. Out of doors there is the same delightful difference. You stride cheerily along and are not harassed at every turn by howling herds of unappreciative trippers'.

It is also worth remembering that this invasion of the hills, not by the 'unappreciative trippers' who rarely leave the roads, but by those who love the hills, is an advantage to the inexperienced. During the usual holiday periods and at most weekends, especially during the summer months, the more popular routes over the fells and the more frequented summits, Skiddaw, Helvellyn, Scafell Pike, Great Gable, Langdale Pikes, Coniston Fells and many others will be traversed by many parties, large and small.

The walker with some experience will soon discover that the well-blazed 'main roads' to the popular fells are not the only walking routes. The ability to read a map and to use a compass when necessary will provide routes and even summits where he is 'not harassed at every turn'. He will soon discover, too, the great enjoyment to be gained from ridge walking as distinct from summit ascending. The district abounds in ridges which can be used to link a series of adjacent summits giving a fell walk of several miles, more or less according to individual desires, on the tops. The very form of the Lake District is that of ridges rather than splendid isolated peaks, a factor contributing to the unity and beauty of the area. Wordsworth early appreciated this in his *Guide to the Lakes*, 1810:

'And first of Mountains. Their forms are endlessly diversified, sweeping easily or boldly in simple majesty, abrupt and precipitous, or soft and elegant. In magnitude and grandeur they are individually inferior to the most celebrated of those in some other parts of the island; but in the combinations which they make, towering above each other, or lifting themselves like the waves of a tumultuous sea, and in the beauty and variety of their surfaces and colours, they are surpassed by none'.

Here no reference will be made to cross-country walks, from dale to dale, which are mentioned in Chapter 10. Most attractive to the mountain lover are the horseshoe circuits of the fells in which the Lake District abounds. These circuits, beginning and ending near the same point, circle the head of a chosen dale along the sky-line and reduce the distance to be walked on roads to the minimum. From Rydal there is a very pleasant walk around the head of Rydale

Fell Walking and Rock Climbing 105

(Rydal Beck Valley), not too long for a winter day, a walk of some 9 miles with a total ascent of not more than 3,000 feet, yet providing 5 or 6 miles of ridge walking above the 2,000-foot level. The route ascends Nab Scar and follows the ridge over Heron Pike and Great Rigg to Fairfield, returning to the starting point over Hart Crag, Dove Crag, High Pike and Low Pike. Five hours would be a reasonable time allowance for an average walker in clear weather, but unless the day is very cold or the weather bad, many would prefer to spend another hour or two in enjoyment of the constantly changing views from the ridge and its many small summits.

Similar is the ridge walk around Oxendale at the head of Langdale, over Bow Fell, ascending by The Band, continuing past Three Tarns, Shelter Crags, Crinkle Crags and Pike o' Blisco (Pike of Blisco on O.S. map) back to Dungeon Ghyll. Those who require a little more exercise can begin, by way of the path to Stickle Tarn, with Pavey Ark and Langdale Pikes, reaching Bow Fell via the summit of Stake Pass, Angle Tarn and Ewer (Ore) Gap. Those who desire a slightly shorter walk can return to Oxendale without including Pike o' Blisco.

Other horseshoe ridge walks can be made over the Kentmere fells, involving Harter Fell in Mardale at the head of Haweswater, High Street, Thornthwaite Crag, Froswick and Ill Bell; or around the head of Mardale, ascending Harter Fell by Gatescarth Pass, then over Mardale Ill Bell to High Street, returning down the long shoulder from Kidsty Pike.

An excellent round from Braithwaite, west of Keswick, includes Grisedale Pike, Eel Crag, Sail and Causey Pike, with Grasmoor added for those who wish. Grasmoor can also be included in the Buttermere round of Whiteless Pike, Wandope (Wanlope on map), Grasmoor, Hobcarton Crag (Hopegill Head) and Whiteside.

One of the most delightful walks in the whole district, more strenuous than the preceding, but within the ability of the average walker accustomed to the fells, is the circuit of Mosedale (Wasdale) over Wasdale Red Pike, preceded by Yewbarrow as an agreeable hors d'oeuvre, continuing to Scoat Fell, Steeple (a short but worthwhile divergence from the main ridge), then over Pillar (the fell, not the crag which is situated on its Ennerdale face) and Kirk Fell with a finish over Great Gable for good measure. A few minutes spent with the Ordnance Survey one-inch Tourist map of the district will show the great number and variety of such walks which can usually be made longer or shorter to suit individual abilities and the prevailing weather.

Fell walking demands simple suitable equipment, notably footwear, and here the choice between nails and moulded rubber soles is left to the individual, although it is worth mentioning that nails have advantages under icy conditions or on mossy rock; moulded rubber is easier to the feet on stony paths. In any event the boots should be strong and waterproof, but not too heavy. Clothing can be left to personal choice, but it is well to remember that the tops can be very cold and stormy, even in summer, emphasising the importance of a windproof jacket and too much, rather than too little, clothing. A lightweight waterproof cape is useful in rain. Certain equipment should always be carried in case of emergency; the essentials are whistle, first-aid outfit, spare food, a lamp especially in winter when days are short, spare pullover, and, of course, map and compass together with the ability to use these.

The number of those walking on the fells, at all seasons, shows a remarkable increase. Unfortunately, in spite of repeated warnings and advice, the number of those venturing there inadequately equipped, especially in winter, is increasing too. The importance of proper boots; warm, windproof and waterproof clothing; map and compass, etc., cannot be too strongly emphasised. Recently there have been far too many instances of walkers becoming benighted on the fells as a result of miscalculating the hours of daylight available in winter. Spending a night out on the fells then is, at best, uncomfortable; in bad weather it can be fatal. It is therefore essential to carry a good electric torch (a spare battery and bulb are advisable) and to plan the route bearing in mind the few hours of daylight and the time which may be lost in route finding. The Lake District has most efficient voluntary rescue teams, ready to come to the assistance of anyone in trouble, but it is incumbent on those going on the fells so to equip themselves and so to plan their routes that the teams are not called out as a result of avoidable mistakes. During a week in winter 1972 one team was called out four times because of walkers benighted; on one occasion 150 searchers were involved. The fells will provide great enjoyment to those who come well equipped and prepared and who learn respect for all mountains, even small ones; they can prove dangerous or fatal to those who ignore the advice of the experienced.

The young and energetic walker, in good training, can plan more strenuous days and there are always a few with ambitions to walk the 3,000-foot summits (Sca Fell, Scafell Pike, Helvellyn and Skiddaw) within 24 hours. This walk as a complete circuit involves some 50 miles of strenuous going and represents an achievement of which any walker may well be proud. Outstanding walkers such as Dr Wakefield, Eustace Thomas and Robert Graham have improved on this walk by adding as many other summits as possible in 24 hours.

Fell Walking and Rock Climbing

The outlying fells may be recommended to those who seek a change of scene and an escape from crowds during the busy holiday periods. Many of these fells have more of the nature of moorlands but without the heavy going associated with peat and Pennines. In addition to solitude and quiet, broken only by the sound of the wind and curlews' cries, an attraction of these walks is the new and often unexpected views of the main fells—a change of scene resulting from a position from which one looks up the dales to the higher fells, rather than down the dales to the surrounding flat lands. Often the best and most colourful views in mountain country are from the mid-heights and not from the high summits, an opinion which can be verified by enjoyable walks over such smaller fells as Mellbreak, west of Crummock Water, and the adjacent Loweswater fells, or Middle Fell and Seatallan from Wasdale Head.

Scrambling

Many variations on the usual pedestrian routes can be made by those who wish for something a little more adventurous without undertaking any actual rock climbing. These routes, intermediate between walking and climbing and sometimes reaching the standard of an easy rock-climb, demand proper footwear, route-finding ability and steady heads. In wintry conditions many of them should be treated as real climbs and may well require the use of rope and ice-axe. Acquaintance with the fells will quickly bring such routes to light; it is possible to mention only one or two here. A walk over Langdale Pikes can be made by following Dungeon Ghyll, keeping close to the stream except where a waterfall requires a detour, then round Stickle Tarn to Pavey Ark which can be ascended by Jack's Rake. This is one of the easiest rock-climbing routes, which means that it is difficult as a walk. On such a route, in wet weather, nailed boots may well prove a happier choice than the more popular type with moulded rubber soles. Anyone with a little rock-climbing experience may complete the day's scramble by descending South-East Gully on Gimmer Crag, another 'easy' rock-climb but considerably harder than Jack's Rake, especially for a novice faced with its descent. Another route, introducing some scrambling, is growing in popularity—the ascent of Great Gable by striking across the Wasdale face from Kern Knotts, near Sty Head, along the climbers' path below the Napes Ridges, then upwards just before the Sphinx Ridge. This is an interesting route but is beset by a mountaineering hazard—loose stones. Great care therefore must be exercised and the route is unsuitable for a large party unless all can be trusted to move safely on loose rock. Accidents have happened here and, among other places, in Pike o'Stickle Gully where there is danger from dislodged boulders. This gully, together with others containing scree or boulders, is a place to avoid when other parties are about.

Scores of delightful routes, some walking, some scrambling, are clearly described in Wainwright's *Pictorial Guides to the Lakeland Fells*.

Rock Climbing

Rock climbing developed a little later than fell walking. Pillar was mentioned in Hutchinson's *History of Cumberland*, 1794, and Otley, although he looked on the crag with obvious approval, described it, in the early editions of his guide, as inaccessible. It gained some notoriety from Wordsworth's poem *The Brothers* (1820) which tells the story of a shepherd who perished in a fall from the crag, although it is doubtful whether in fact the accident occurred on Pillar. Probably the first authentic account of a Lake District rock-climbing ascent was that of Pillar by John Atkinson in 1826, a feat recorded in the local Press. There seems to be no doubt, however, that years earlier, in 1802, Coleridge descended from Sca Fell to Mickledore by the route now known as Broad Stand. Although Coleridge was often inexact in the names he gave to the fells, there is a remarkably accurate description of this route in the Journal of his tour made in 1802. Another early reference is found in the *Penny Magazine* of 1837 to a way known to shepherds of reaching Sca Fell from Mickledore.

Other Ennerdale shepherds also climbed Pillar in the early years of the last century but, apart from the descent of Broad Stand by Coleridge, there is no record of any climb by a tourist until a Lieut. Wilson climbed Pillar in 1848, an event which may be regarded as the first sporting rock-climb in the Lake District. During the second half of the century rock climbing developed slowly, with new routes on Sca Fell and Pillar. In 1870 ladies enter the scene; Miss A. Barker climbed Pillar, a climb repeated by Miss May Westmorland with her brothers in 1873. The son of one of these brothers was still to be seen on the fells in 1974.

The year 1886 was marked by an important event—the first ascent of the Napes Needle—a solo climb by Haskett Smith, to whom the title of father of English rock climbing is sometimes given. It would be more accurate, however, to refer to him as one of the fathers, for we must not forget Seatree who was climbing in 1865 and Robinson who climbed Pillar in 1882 and who made over one hundred ascents of that crag. His is a unique memorial—the High Level Route from

Looking Stead, above Black Sail Pass, to Pillar, with the Robinson cairn at the end of the route when Pillar comes into full view. Again there was the Rev. James Jackson, 'The Patriarch of the Pillar', who made his last ascent of 'his' crag in his 80th year. To return to Haskett Smith: he exercised a great influence on the sport of rock climbing in the eighties, in the period when climbers began to forsake the easy ways to the summit of a crag for the more difficult routes in gullies and chimneys. His little volume *Climbing in the British Isles, England*, 1894, was the first rock-climbing guide-book. In 1936 this great climber was able to make his Jubilee ascent of the Needle, at the age of 75.

This second half of the nineteenth century saw many new routes opened up by alpinists who visited the Lake District in winter, for climbing practice and exercise, and who were attracted by snow-filled gullies. Then, following Haskett Smith's example, climbers began to accept the challenge of steep difficult rock and left the recesses of gully and chimney, with their impression of security, for the greater exhilaration of 'outside' climbing. The reader who is interested in these developments may refer to the series of climbing guide-books published by the Fell and Rock Climbing Club, previously under the editorship of H. M. Kelly: a new and up-to-date series of eight volumes, edited by John Wilkinson, has recently been published (see Bibliography, page 139).

As long ago as 1887 Hopkinson gave a foretaste of things to come in a December attempt on the Pinnacle Face of Sca Fell. The attempt was defeated by ice on the upper part of the face from whence a safe retreat was made. This attempt was a formidable and outstanding undertaking in the gully-climbing era, long before climbers had begun to realise the possibilities of such exposed and delicate climbing; indeed it was another quarter of a century before Hopkinson's route was completed. Two others who made remarkable advances in this type of climbing during the last decade of the century were Solly and O. G. Jones. The former led Eagle's Nest Ridge (direct) on Great Gable in 1892. The ridge is steep and exposed, many of the holds are small and sloping, and the climb is still rated as 'severe' and treated with respect. In 1896 Jones, accompanied by the Abraham brothers, led Scafell Pinnacle from Deep Ghyll and, two years later, led the still more exposed and severe route direct from Lord's Rake.

With the advantages of developments in technique and a background of the accumulated experience of their forerunners, climbers sought more and more difficult problems. Thus, in the early years of this century, interest turned to new climbs, often more exposed and delicate, on slabs and walls. Routes thought at one time to be of outstanding severity were climbed ever more frequently and became the normal rather than the exceptional. This intensification of progress has continued, experience has accumulated and, mindful of the adage 'What man has done, man can do', climbers developed their individual skill more rapidly. Equipment improved; the introduction of 'rubbers' (gym shoes) permitted a much higher standard of climbing, on dry rock, than was possible with standard nailed boots, while improvements in transport made it possible to spend frequent weekends and also more hours of a climbing day on the crags. A factor of special note was the effect of gritstone climbing on rock climbing in general. Climbers in the industrial towns, particularly those near to the Pennines, sought outlets for practice close at hand. The gritstone outcrops of the Pennines provided an answer. These outcrops, although relatively low in height, demand a technique based on balance rather than on muscular effort. This balance technique, developed on the low crags, made itself felt on the higher crags and a general raising of the standard of what is possible followed.

Periodically rock-climbing progress appears to reach the limit of the possible, but before long another step forward is made and a new standard reached. Such an event was the successful climb of Central Buttress on Sca Fell by Herford, Sansom and Holland in 1914. It was seven years before this climb was repeated, which, even making allowance for the war years, indicates its severity. In the present climbing era Central Buttress, although treated with great respect, is climbed by several parties each year. Other similar steps forward, among many which must pass unrecorded here, are exemplified by Kirkus's lead of Mickledore Grooves on the Eastern Buttress of Sca Fell, in the early thirties. This was followed by other still harder climbs on this very steep and severe buttress, led by Linnell.

The detailed recording of outstanding events since the early thirties, within the narrow limits of this chapter, becomes quite impossible, therefore this historical survey of rock climbing must be brought to a close by a brief mention of recent trends. The ever-growing increase in the number of climbers is worthy of special note. In the first place it produces a rapidly growing band of first-class leaders who are bound to affect the future developments in climbing at home no less than the future of mountaineering and exploration in general; many are men who will find an outlet for their skill and natural sense of adventure in the greater mountain ranges, and in polar regions.

On the other hand, among this large body of climbers, are many not yet familiar with the traditions of mountaineering or even of country behaviour in general. It is of the utmost importance that the goodwill which has for long existed between climbers and other visitors to the district on the one hand, and farmers, landowners and the local people on the other, should not suffer as a consequence.

Another trend has been a remarkable increase in the number of climbing routes and the opening up of new crags. Two examples may be of interest, both post-war. The first was the remarkable 'discovery' of Shepherds' Crag in Borrowdale at the head of Derwent Water by Bentley Beetham in 1945. This crag, within ten minutes' walk of the Borrowdale highway, had been passed by thousands of climbers who failed to realise its possibilities. It now provides some 40 routes, easy to very severe, and is a wonderful practice ground, often affording pleasant climbing when bad weather makes the higher crags impracticable. The second recent development has taken place on various crags in the eastern part of the district, on Raven Crag at the foot of Thirlmere, the Castle Rock of Triermain on the other side of the lake, and on the crags of Deepdale and Dovedale to the south-east. Some idea of the achievements here may be gathered from the guide-book to this area, which now lists over one hundred new climbs in the last decade compared with 64 in the previous sixty years. Of these hundred new climbs 39 are rated as 'very severe'.

Lest anyone should be rash enough to make predictions about the future, it may be salutary to quote from George Abraham, *Complete Mountaineer*, 1907:

'Nowadays it is extremely difficult to find a new feasible route on any of the bigger crags... It is no exaggeration to say that some of the British climbs have reached the border line where the law of gravity exerts itself and human muscles and endurance cannot prevail.'

Rock climbing requires equipment basically similar to that of the fell walker, with the addition of a rope, usually nylon. A further, and most essential, requirement is the knowledge of the proper use of a climbing rope. This technique, which removes so much of the apparent danger from climbing, is best learned at the hand of an experienced climber, either a qualified Lake District guide or an amateur, who will also instruct the novice in the proper use of whatever hand and foot holds are available and also in the balance method of climbing, essential on harder, more exposed climbs and advantageous even on the less severe, to reduce the physical effort required. That rock climbing is a question of balanced movement more often than of muscular effort is demonstrated by the great skill and ability of many women climbers. Although skilled instruction in climbing is desirable there are novices who begin with equally inexperienced friends. This is not unreasonable if restraint is exercised in the early days, and particularly if such climbers have an occasional opportunity to follow an experienced leader on harder climbs. Finally, physical fitness is required for rock climbing; fitness and safety go together, especially where the leader is concerned. There is therefore a good deal to be said for beginning a climbing holiday with a day or two of fell walking for anyone not in good training.

It is not possible to give here detailed descriptions of the hundreds of climbing routes available, nor even to name all the crags on which climbing is now enjoyed. These details are given in the guide-books to which reference has already been made. Those unfamiliar with the district, and also those who do not climb, might be interested to know where the main crags are situated. The traditional climbing grounds are Dow Crag (Coniston), Scafell Crag, the Napes Ridges and Gable Crag on Great Gable, Pillar (Ennerdale), Pavey Ark and Gimmer Crag (Langdale) and the crags of Birkness Combe overlooking Buttermere. Much climbing is also done on various crags in Borrowdale—Comb Gill (Glaramara), Gillercomb above Seathwaite, Shepherds' Crag near Lodore—and also in Deepdale and Dovedale near Brothers Water and in perhaps another dozen localities within the National Park. These crags provide routes suitable for climbers of all degrees of skill and experience. On each crag certain climbs stand out: routes not necessarily of great difficulty, but having particular charm either because they provide a natural line from base to summit of the crag or because they involve situations and moves which give great satisfaction to the climber. Some natural routes are comparatively easy and thus provide excellent training climbs for parties of limited experience; such routes are C and D on Dow Crag, Arrowhead Easy and Needle Ridge (not the Needle itself) on Great Gable. Others are of a higher standard of difficulty and include such classical courses as Gordon and Craig's Route on Dow Crag, the New West on Pillar, Tophet Bastion on Great Gable, Great Gully on Pavey Ark, Moss Ghyll on Sca Fell, and that particular Lake District gem—the Napes Needle. A third group of severe climbs, requiring a leader of some ability and experience, is typified by Eagle's Nest Direct on Great Gable, the Rib and Slab on Pillar, and Central Chimney on Dow Crag. This group merges, through still harder climbs such as Tophet Wall on Great Gable, the Crack on Gimmer, and Central Route on Bow Fell Buttress, into the higher-groups—the 'Very Severes' and the 'Extremely Severes'. This last group, a development of the past 15 years, represents the present limit of climbing ability of the very few.

Progress has been rapid largely because of the increased opportunities to climb afforded by more leisure time (if leisure can be applied to rock climbing) and the ease of reaching the Lake District. A young rock climber can now, in his first year, reach a much higher standard than was formerly attained after several years. Perhaps even more striking have been the effects of improvement in equipment and technique. The one-time traditional nailed boot has been almost universally replaced by Vibram-soled boots, while very light, close-fitting boots (generally known to climbers as P.A.s) are carried and used on the harder climbs. Nylon ropes, adopted for 20 years or more as standard equipment instead of the old hemp ropes, afford a much greater margin of safety, especially for the leader. Outstanding progress has been made in methods of protection by means of running belays which reduce the risk to the leader by minimising the extent of a possible fall, thus enabling harder climbs to be accomplished without increased risk. One looks forward with great interest to the next decade.

No mention has been made in this chapter of the best centres for walking or climbing. Much depends on the individual's choice and, if motor transport is available, the choice is very wide as the Lake District National Park is a compact area in which the part of interest to a mountaineer can be circumscribed by a circle of no more than 15 miles radius. It is possible to stay in Keswick and walk the fells in the immediate neighbourhood and climb on the local crags, or alternatively, from the same headquarters, visit fells or crags as widespread as High Street, Wetherlam and Pillar (both fell and rock), or Dow Crag and the Deepdale climbs. Many will think it more agreeable to select a centre such as Borrowdale, Langdale, Patterdale, Wasdale or Buttermere, to mention but a few, from any of which a fortnight's walking or climbing is possible without the use of road transport. Some centres may be better than others, some fells pleasanter than others, some crags to be preferred to others, but all are good. (See also Figure 8.)

12
Traditional Sports

by A. H. GRIFFIN

LONG before the Lake District began to attract its first tourists, when it was still a rather wild and remote area, the dalesmen were wrestling on the village greens and chasing the fox across the fells. It was part of their way of life—their heritage from the hills—and remains so today. For you need strong legs and good lungs to hunt the fox, and strong arms and good balance for the wrestling, and the mountains breed such men.

These two, fox hunting and wrestling, are perhaps the oldest of the traditional sports of the fell country, but there was hound trailing and fell racing at the very first Grasmere Sports much more than 100 years ago, and some other Lakeland sports are old enough by now to have become traditional.

Note the mountain association in three of the four principal traditional sports, while wrestling, today held against a backcloth of the fells, at one time took place on the summit of High Street. For the dalesman has little interest outside the fells and is not even greatly attracted to the lakes themselves.

The Royal Windermere Yacht Club was 100 years old in 1960, there have been motor boats on the same lake for more than 60 years, and today there are the swimming races and the water skiing, but these things are not bred in the dalesman. One can imagine him looking on, with some wonder and amusement, at the regattas on Derwent Water and Windermere more than 150 years ago, but almost begrudging the time spent away from the hunting or the wrestling. For, while you can sail and swim wherever there's sufficient water, it should be realised that hunting the fox on foot, Cumberland and Westmorland style wrestling, hound trailing and fell racing, although tending to spread their tentacles nowadays, are all sports that were born in the fell country and are rarely seen outside it.

There is little in common between fox hunting in the shires and following the fell packs in Lakeland, for the dalesman hunts on foot and only the huntsman wears the red coat. You may ride to hounds with a fashionable hunt because it is the thing to do or for sport and exercise, but in the fell country there are no social niceties, for killing foxes is just another job like keeping down the crows or repairing fences. For the Lake District fox, in his long nightly prowls for food, slinks down from the crags and the upland slopes to the farms in the dale to prey on the poultry and the young lambs, and the farmers and shepherds who make up most of the hunt are out for his blood.

But they are out for sport as well, for the real dalesman is born to the hunt. Sometimes he is out soon after dawn and a kill before breakfast, before the sun has ruined the scent, is common enough. There are six mountain packs—the Blencathra (sometimes called the 'John Peel' hunt) hunting the Keswick country, the Eskdale and Ennerdale whose territory is the wild western dales, the Ullswater who hunt the Patterdale and Mardale fells, the Mellbreak (Buttermere and Loweswater), the Coniston (the Furness fells and southern Lakeland) and the Lunesdale who operate over a wide territory outside the Park to the east of the Shap Fells road.

The huntsman most associated in many parts of the world with Lake District hunting is the legendary John Peel, but in fact his hunting country was largely outside what is now the National Park, and he was a flamboyant character rather than an heroic figure or even a particularly great sportsman. We remember him today, not for what he was, but rather because of the song written by his old friend John Woodcock Graves and, more often than not, sung outside Cumberland and Westmorland with the wrong words. For his coat was not 'so gay' but 'grey', he lived at Caldbeck and not Troutbeck, while the line 'Ranter and Ringwood, Bellman and True' is quite wrong. No Lake District hound has ever had a one-syllable name, and the correct version, slightly revised from Graves' original dialect, is 'Yes, I ken John Peel and auld Ruby, too, Ranter and Royal and Bellman, as true'.

Peel hunted the country at 'the back o' Skiddaw' and towards Carlisle, and, unlike the mountain pack huntsmen of today, he often hunted on horseback—or rather on the back of what was probably a fell pony (although sometimes described as a Galloway). He was a big, rough man with a loud raucous voice and perhaps some unpleasant habits, but he loved dogs, and hounds in particular. His 'coat so grey' was a long one with brass buttons and he generally wore corduroy knee-breeches, long stockings, and a tall, weather-beaten, beaver hat. In his right hand he carried his whip and his battered, curved copper horn. He had long legs, a great nose, piercing blue eyes and a 'Tally ho' which, they said, could be heard for miles. He would be out at daybreak after a heavy night's drinking and some of his hunts covered up to 70 miles. One day in 1812 he is said to have hunted the same fox for ten hours and forty minutes after which, I am glad to say, the fox escaped. In his prime he kept and hunted 14 couples of hounds and he had a pack of sorts for something like 55 years, but his records of kills in a season have often been exceeded. When he died in 1854 at the age of 78 some of his hounds were acquired by Sir Wilfred Lawson of Brayton Hall, but others went to the Blencathra, which was just being founded about that time by the great sportsman John Crozier, the squire of Threlkeld. And these hounds were used for breeding the forbears of the present pack.

But we hear little today of other fell country huntsmen who in some ways were perhaps even more successful than John Peel and probably finer men. It is more than 100 years since the Eskdale and Ennerdale was founded by little Tommy Dobson who was once a bobbin turner by trade. More than once Tommy went without his dinner so that his hounds and terriers did not go short, but he hunted the wildest country in Lakeland with resounding success. He was probably responsible for the destruction of more Lake District foxes than anybody else up to his time. Tommy must have been a mighty walker for often he had to cross into a distant valley just to collect his hounds and he would tramp right across the Lake District to look at an interesting litter of terriers. Before he died he directed that all his hounds and terriers should be given to Billy Porter, the farmer's boy who, from the age of four or five, had sat at his feet and learned the hunting game; and W. C. Porter succeeded him as Master, reigning for 42 years until his death in 1952. When Tommy Dobson died in 1910 they carried his body in a farmer's trap over the passes into Eskdale for burial and the dalesfolk knew that a great character had passed from their midst. And there have been other mighty hunters besides Crozier, Dobson and Porter. There were the Rev. E. M. Reynolds who carried on the Coniston hounds for a generation around the turn of the century, the great Joe Bowman of the Ullswater who died at Glenridding in 1940 at the age of 90, and Joe Wear, the recently retired huntsman of the same pack who probably killed more foxes than John Peel or anybody else. Anthony Chapman, huntsman of the Coniston pack, still continues an old family tradition.

To the dalesman hunting means the thrill of the chase, often over the roughest country in England, but there is also, deep down in his breeding and traditions, the love of a man for a dog. The hounds, which lead an easy life on the farms during the summer, are all known to him by name and he is a proud man indeed when the animal he has trained or 'walked' is adjudged a champion at one of the summer hound shows in the dales. For the fierce little terriers sent into a borran or hole among the rocks to bolt the fox he has a very real affection and if the terrier is trapped underground, as often happens, he will leave his job and spend days with spades and crowbars trying to dig him out. For the follower, hunting is a hard, lung-racking game, for the fox nearly always makes for the heights and you need a knowledge of the ways of a fox and something of a feeling for the lie of mountain country, besides sound limbs and a good heart, to be up with the kill. To some of us it may seem a sickening business, this chasing of a beautiful animal to the death. But to set against this apparent cruelty there is the necessity to keep down the numbers of foxes (which are also shot by the pest destruction societies) and the knowledge that the fox is normally killed in an instant by a snap from the leading hound. And the pack does not break up its fox in Lakeland nor is he deprived of his brush, but more often hung up in state upon a farmhouse door—an honoured foe and perhaps an example to his fellows.

When the hounds are resting from the hard work of the winter the dalesman turns to his summer preoccupation with dogs—the north country sport of hound trailing. Trail-hounds are basically foxhounds but are specially bred and trained for a different purpose. Hound trailing can provide some fine spectacles—the excitement of the slip, the wonderful sight of distant specks coursing along a shoulder of fell, and the thrill of a close-fought finish—but undoubtedly the betting is the principal attraction. The sport attracts not only the dalesmen, but also the miners from west Cumbria, the shipyard workers from the Barrow area, and artisans and business folk from towns beyond the fringes of the National Park. Every agricultural show, every sports meeting and sheepdog trial has its hound trail—generally two or three in an afternoon—and in addition there are trails on nearly every day of the week in one part of the district or another. The whole business is highly organised with a ruling body founded as long ago as 1906 and the corruption which was formerly a feature of the sport is now rarely attempted. There was a time when hounds were substituted or drugged or tricked into drinking or even taken along part of the route by car, but severe penalties, strict supervision and closely guarded trails have made cheating not worth the candle.

Two men dragging old stockings filled with an official mixture strongly flavoured with aniseed lay the trail, the men starting off from a point halfway round the course and approaching the start and finish by a different route. Watch the finish of a hound trail and, whether or not you know anything about the game, you will catch the thrill. There has been no sign of the hounds for perhaps a quarter of an hour, but then somebody catches a sight of the leaders through his glasses. A tremor of excitement runs through the crowd and in a few moments we can see them with the naked eye and the catchers, standing in front of the spectators, set up a shrill chorus of whistles and cries. The leaders jump the last few walls and make straight down the field towards the call they know and, seconds after crossing the line, they are gulping meat from a tin bowl—their reward for the long struggle across the fells. They are groomed and cosseted like racehorses and although doubtless their owners show them affection they are really the stock-in-trade of the itinerant bookie and a symbol of easy money for the punter. What could be a fascinating outdoor spectacle is really no more than a gamble.

But wrestling—wrestling in the Cumberland and Westmorland style—has little or no attraction for the betting man, and although there are tales of faked contests in the past it is nowadays one of the cleanest of sports. The best man wins and the man who throws all his opponents, working his way through several rounds to the final, becomes the champion of the meeting at his weight. And each year the best wrestlers at particular competitions, listed in advance by the ruling body, become the champions of the world in this particular style—an honour which many an 18-year-old farm lad from the dales may rightly claim.

The rules are simple. You stand with your arms locked behind your opponent's back and the contest ends when one man is on the ground; if both fall together the winner is the one on top. In the early stages of the contest the men slowly circle the grass, heads on one another's shoulders and legs well apart, seeking an opening or a sign of relaxed concentration on the part of their opponent. Then comes the attack, with a great heaving and tripping, and the round may be over in a trice, or they will recover and battle on. The spectator may not realise it, but the game is highly technical and there is much to learn, the secret being to tempt your opponent into a position of apparent security and then quickly to get him off balance. Mere strength is by no means the whole of it but the best men at the game seem well built and sturdy, the heavyweights often large indeed (Plate XII).

Perhaps the outstanding man in the long history of Cumberland and Westmorland wrestling was the great George Steadman who died at Brough in 1904. A photograph taken in his prime shows a powerful but paunchy man with a benign smiling face not unlike a bishop's, a great bald, domed head, the white side-whiskers of a patriarch, and an easy confident stance. He is dressed in white vest and trunks, eight medals hang from a chain around his neck and round his tremendous waist is clasped a massive, ornate belt. At his side and filling most of the picture is a stand containing enough cups, tankards, rose bowls and medals to stock a fair-sized jeweller's shop.

Steadman won the heavyweight championship at Grasmere on 14 occasions, was England's representative in several international contests in London and abroad, and probably made more money at the game than any other man, before or since. His chest measured $51\frac{1}{2}$ inches—his waist might have been even bigger—and his calves nearly one foot six round. But despite these cumbersome proportions he could be as nimble as a dancer when he chose and it is said he was a gentle man with children and animals. Yet, when more than 50 years of age, he won a wrestling contest in Paris against ten Englishmen, seven Frenchmen, six Swedes, four Australians, two Italians, two Americans and one Greek, disposing of each of his opponents in turn. Ted Dunglinson of Carlisle has improved on Steadman's total of Grasmere wins, but the old patriarch remains the personality of the sport.

Wrestling in the dales is a sport for the man who knows the game and not really for the casual spectator who cannot appreciate its finer points, but for those who can sense atmosphere it has something of an old-world flavour—of Merry England perhaps—and even of beauty. The sunburned wrestlers in white hose and gaily embroidered trunks make a fine picture as they slowly circle the close-cropped turf below the ring of fells, and the handshake at the end highlights the friendliness of it all.

In the winter the wrestlers carry on their sport indoors—academy wrestling it is called—but the great championships are won at Grasmere and the other principal dales sports and the present-day champions often bear the same names as the heroes of a hundred years ago. With so many counter-attractions for the young dalesman there was a sign a few years back of declining interest, but old traditions die hard and a revival is under way. Sports promoters are co-operating by insisting on the wearing of traditional costume; sloppy socks and braces are going out. Wrestling is one of the finest heritages of the dalesman—a simple manly sport, unsullied by gambling, where a young man may glory in his strength and skill.

The beginnings of fell racing in the Lake District have long since been forgotten, but there was a fell race—or guides' race, as it is sometimes called—at Grasmere in 1852, the first meeting for which records still remain. To many people the fell races are the principal attraction of the dales sports—Grasmere, Ambleside, Patterdale and the rest—for this is sheer spectacle, a race to the top of the nearest fell and back again. We see them, high up on the heights, rounding the flag at the summit after their gruelling struggle up the fellside and then, in a few minutes, after their breakneck dash down through the bracken, over walls and through the rocky outcrops, they are racing down into the arena while the band strikes up 'See the conquering hero comes'.

In recent years Billy Teasdale, a short, wiry dalesman who farms in the John Peel country, has been the outstanding fell runner, but in the great days of George Steadman, Jack Greenop had six Grasmere victories up to 1881. Then there were Jim Grisedale of Braithwaite—a nursery of many famous runners—Tom Conchie, the great Ernie Dalzell of Keswick who won six years running at Grasmere and then, after two defeats, returned to win again, George Woolcock of Hawkshead, Ronnie Robinson of Spark Bridge, Ronnie Gilpin of Braithwaite and Stanley Edmondson of Borrowdale.

An old photograph of the Grasmere runners in 1875 shows two of the eight competitors wearing low shoes and all of them, except one, wearing what look very like singlet and long underpants, with bathing trunks added. The single exception wears heavy, corduroy knee-breeches, a pair of braces, and a powerful beard.

Here is another sport where the bookmaker has intruded, but the game is clean enough and the honour of winning at an important meeting still highly prized. Nowadays the sport has spread outside the limits of the National Park, but it was born in the fell country and the best men of all are young farm lads and shepherds to whom the peak of physical fitness presents a worthwhile challenge. Even youngsters of ten and twelve have their races and from them come the great champions of later years.

Pole leaping is another tradition at Grasmere, while the old Crab Fair sports at Egremont have their gurning—or grinning—competition. You place your head through a horse collar and the winner is the man who pulls the ugliest face. Norman Nicholson quotes a delightful story of a former champion gurner of whom it was said that the first time he won the prize he had not really entered the competition 'but was merely following the efforts of the others with interest and sympathy'.

These then are the principal traditional sports of the fell country, but I am tempted to mention also the sheepdog trials which, although nowadays held in many parts of Britain and even abroad, had their beginnings in the north, and were early fostered in Lakeland. The very first English sheepdog trials took place in Northumberland, not far south of the Border, in 1876, and in the following year they were held on Belle Isle, Windermere 'on the last Saturday in August and before a large and fashionable gathering'. Regular trials began to take place in the northern counties, the Lake District Sheep Dog Trials Association was formed in 1891 and, apart from war years, these trials have continued annually at Applethwaite near Windermere ever since. Trials began at Rydal in 1901 and Patterdale's 'Dog Day' has a long history.

Basically, the sheepdogs are the Border Collie, the friendly black and white animal with the silky coat and lively eye you will find on almost any Lakeland farm, and you could never meet a more intelligent creature. There is not a sound from the watching thousands as the dog slowly advances towards the sheep for the 'pen', sometimes lying full length on the ground and edging forward an inch at a time, sometimes creeping softly behind them, but never for a moment taking his eyes off them. He is, in a sense, hypnotising the sheep, forcing them with his glinting eye to do his bidding. The tension in an important trial, when perhaps men and dogs are being selected to represent their country, can be great, for the slightest movement from frightened sheep when time is running out but success seems near can snatch victory out of the shepherd's grasp. But neither the shepherd nor his dog must show exasperation or disappointment, and some of the best handlers are the most patient and kindly of men. It is men like this who have helped to make the Border Collie the splendid animal it is today, for the one deserves the other.

You can watch these trials, held in the most glorious surroundings throughout many a long summer's day, and the dales sports, almost every summer weekend. But mark the date of the most important meeting—Grasmere, on the Thursday nearest to 20 August. Here you will see the traditional sports of the Lake District at their best, amid the most wonderful scenery in England.

13
Literary Associations

by GEORGE BOTT

WHAT *was* the first piece of Lakeland literature? Does a biography of St. Kentigern written in 1180 by Jocelyn, a monk of Furness Abbey, qualify? The location may have only a fringe claim but Kentigern's missionary journey through the Lakes in the sixth century is signposted by churches dedicated to him at Caldbeck, Mungrisdale, Keswick and elsewhere.

Or perhaps Richard Braithwaite of Burneside merits the honour? His account of the ferry disaster on Windermere in 1635 written in the following year reports an accident in which 47 people were drowned. It has all the drama of tragedy: the occasion was 'nuptiall but fatall'; the boat, heavy with horses and carriages, 'became drench'd in the depths' and 'not one person of all the number was saved: Amongst which, the Bride's Mother, and her Brother in this liquid regiment, equally perished'.

Or what of the anonymous ballad of *Adam Bell, Clym o' the Clough and William of Cloudeslie*, a stirring tale of northern Robin Hoods in the forest of Inglewood?

The question remains academic. By the eighteenth century there is no doubt that Lakeland literature had been launched. This was the age of the curious traveller, viewing mountains at a distance and displaying conventional terror at the monstrous hills and impending crags. Dr John Brown looked at Derwent Water in 1766 and saw 'rocks and cliffs of stupendous heights, hanging broken over the lake in horrible grandeur, some of them a thousand feet high, the woods climbing up their steep and shaggy sides, where mortal foot never yet approached'.

The early travellers produced a rich crop of diaries and journals —Gilpin and West, Fiennes and Defoe, Pennant and Budworth. Thomas Gray, best known for his *Elegy Written in a Country Churchyard*, came north in 1769; his companion, Dr Wharton, was taken ill at Brough with a violent fit of asthma and Gray continued his tour alone, sending his sick friend a daily account of his doings. He can be factual: Dalemain, near Pooley Bridge, is 'a large fabric of pale red stone, with nine windows on the front and seven on the side'. He can be poetic: he walked by Derwent Water 'after sunset, and saw the solemn colouring of the night draw on, the last gleam of sunshine fading away on the hill tops, the deep serene of the waters, and the long shadows of the mountains thrown across them, till they nearly touched the hithermost shore'. He can be absurdly romantic: he ventured no farther than Grange-in-Borrowdale and accepted without question the reported dangers and difficulties of the path over Sty Head, tramped today by thousands of visitors from plimsoll-shod toddlers to game grannies, but for Gray:

'All farther access is here barred to prying mortals, only there is a little path winding over the fells, and for some weeks in the year passable to the dalesmen; but the mountains know well that these innocent people will not reveal the mysteries of their ancient kingdom, "the reign of *Chaos* and *Old Night*", only I learned that this dreadful road, divided again, leads one branch to Ravenglass, and the other to Hawkshead.'

The mysteries hidden from prying mortals might well have been smugglers' loot or the odd illicit still!

Mrs Ann Radcliffe, nurtured on the imaginative flights of her own novels, tackled Skiddaw in 1794. She edged her way along 'the brink of a chasm' and narrow precipices; she found sublime streams hurrying into the abyss and 'to save ourselves from following, we recoil from the view with involuntary horror'; as she ascended, the air became thin, 'boisterous, intensely cold, and difficult to be inspired', and her Alpine struggles are neatly completed by meeting on the summit a local farmer on his first climb of the mountain, 'so laborious is the enterprise reckoned'.

It was a Cumbrian who injected common sense and proportion into Lakeland literature and began the real literary associations of the district. William Wordsworth, born in Cockermouth in 1770 and educated at Hawkshead, lived for most of his life in the Grasmere–Rydal area. However uneven his poetry may be—and it ranges from the sublime to the banal—his reputation was a magnet that drew his fellow-craftsmen inexorably to the Lakes, and as writers they crystallised their impressions in verse and prose that echoes their distinctive personalities.

Of Wordsworth and his sister, Dorothy, so much could be said. Their devoted disciples still quietly explore the haunts that were dear to both brother and sister, while thousands renew their brief association, so often lamentably restricted to learning 'I wandered lonely as a cloud' at school, by a conducted tour of Dove Cottage. The real Wordsworth is enshrined in the fells and becks, the landscape he explored on foot. De Quincey was in no doubt about Wordsworth's prowess as a walker, even though:

'His legs were pointedly condemned by all female connoisseurs in legs ... I calculate, upon good data, that with these identical legs Wordsworth must have traversed a distance of 175,000 to 180,000 English miles—a mode of exertion which, to him, stood in the stead of alcohol and all other stimulants whatsoever to the animal spirits; to which, indeed, he was indebted for a life of unclouded happiness, and we for much of what is most excellent in his writings.'

The sheer bulk of Wordsworth's poetry demands rigorous selection, an essential exercise to separate the voice of the deep from that of an old half-witted sheep, to use J. K. Stephen's paradox. Many of the shorter poems, much of *The Prelude*, some of the philosophical musings speak trumpet-tongued of a poet rated by Matthew Arnold as third in line to Shakespeare and Milton. Much of his prose is now ignored but his *Guide to the Lakes* (to use the simplest and most direct of the various titles it had) still remains a balanced and penetrating survey. It has its moments of picturesque jargon and romantic blur but the combination of accurate observation and informed insight points from the earlier effusions of his predecessors to the more reliable guide-books of the nineteenth and twentieth centuries.

If we still read Wordsworth's 'Companion for the *Minds* of Persons of taste, and feeling for Landscape, who might be inclined to explore the District of the Lakes with that degree of attention to which its beauty may fairly lay claim', we should equally be aware of Dorothy's *Journals*, for here, too, a poet speaks. Coleridge commented on 'her eye watchful in minutest observation of nature' and William himself gratefully acknowledged that 'she gave me eyes, she gave me ears'. Dorothy records the minutiae of everyday life— 'I baked pies and bread'—alongside rare moments of intensity:

'There was something in the air that compelled me to serious thought—the hills were large, closed in by the sky ... the moon came out from behind a mountain mass of black clouds. O, the unutterable darkness of the sky, and the earth below the moon! and the glorious brightness of the moon itself! There was a vivid sparkling streak of light at this end of Rydale water, but the rest was very dark, and Loughrigg Fell and Silver How were white and bright, as if they were covered with hoar frost.'

When the Wordsworths settled at Grasmere in 1799, Coleridge— a perverse but genuine friend—lost no time in moving to Greta Hall in Keswick. His career, like his later relations with William and Dorothy, was as unpredictable as Lakeland weather: a sad, disturbing story of misunderstandings and unfulfilled hopes. Coleridge's brief stay at Greta Hall was the prelude to the 40 years his brother-in-law, Robert Southey, spent there, slaving with the application and industry of a committed artist, producing vast quantities of verse and prose. Famous and admired in his own day, Southey is remembered for his *Life of Nelson*, a handful of poems, and the nursery tale of *Goldilocks and the Three Bears*.

Dove Cottage (and later Rydal Mount) and Greta Hall attracted hosts of friends and admirers. De Quincey, recognising *Lyrical Ballads* as 'the greatest event in the unfolding of my mind', twice went as far as Coniston before he dared to meet Wordsworth face to face in 1807; the acquaintance blossomed and prompted De Quincey's *Recollections of the Lakes and the Lake Poets*, a miscellany of biography and criticism, topography and gossip, that has lost none of its interest and value over the years.

Charles Lamb, attached like a limpet to London and 'not romance-bit about Nature', was persuaded to venture north to Keswick in 1802. He and his sister, Mary, visited Ambleside and Ullswater and climbed Skiddaw; Lamb himself waded up the bed of Lodore Falls. A letter to his friend, Manning, sparkles with praise of Lakeland: 'We thought we had got into fairy-land', he admits, and writing to Coleridge he confesses: 'I shall remember your mountains to the last day that I live. They haunt me perpetually'.

Hazlitt was also a guest at Greta Hall. Although the evidence is by no means proven, he appears to have treated one of the local girls rather shabbily and left under a cloud. Shelley spent three months at a cottage overlooking Keswick with Harriet Westbrook, his 16-year-old bride; he was vastly impressed with the view—'the scenery here is awfully beautiful'—but his efforts to win Southey's confidence were far from successful. His pistol practice in the garden and explosive chemical experiments did not endear him to either his neighbours or his landlord and he was asked to leave the cottage. 'We felt regret at leaving Keswick', he writes. 'I passed Southey's house without *one* sting. He is a man who *may* be amiable in his private character, stained and false as is his public one'. John Keats saw his first waterfall at Ambleside in 1818; following in Lamb's footsteps, he scrambled up Lodore and climbed to the top of Skiddaw (with the help of two glasses of rum 'mixed, mind ye, with Mountain water'). He admired Castlerigg stone circle outside Keswick and may well have recalled the visit in a simile in *Hyperion*:

'... like a dismal cirque
Of Druid stones, upon a forlorn moor ...'.

A much more energetic member of the Wordsworth circle was John Wilson, later Christopher North of *Blackwood's Magazine*, who married Jane Penny of Ambleside. He came into the Lakes society like a whirlwind, sailing his boats on Windermere, wrestling with the local lads, indulging a passion for cock fighting. De Quincey tells a story of meeting Wilson on White Moss Common at dawn one summer's morning: he and two companions were on horseback and, 'armed with immense spears fourteen feet long', were chasing a bull—an unusual pastime for a man who became Professor of Moral Philosophy and Political Economy at Edinburgh University.

In 1825 Wilson organised a regatta in honour of Sir Walter Scott; among the guests were Canning, Lockhart and Wordsworth. Scott had stayed with Wordsworth at Dove Cottage 20 years earlier, slipping away to the White Swan for a dram and climbing Helvellyn to visit the scene of Charles Gough's death near Red Tarn. Later he was to use Castle Rock at the foot of St. John's in the Vale in his *Bridal of Triermain*:

> '... midmost of the vale, a mound
> Arose, with airy turrets crowned,
> Buttress, and rampires circling bound,
> And mighty keep and tower.'

There can be few literary celebrities of the nineteenth century without some Lakeland association, however tenuous. Anthony Trollope lived for a time in Penrith; Felicia Hemans, who wrote 'The boy stood on the burning deck', moved to Dove Nest overlooking Windermere, very near to the farm used by Stanley Weyman in *Starvecrow Farm* and a mile or so away from The Briery, the home of Sir James Kay-Shuttleworth, where Charlotte Brontë first met her future biographer, Mrs Gaskell; Dickens and Wilkie Collins climbed Carrock Fell in 1857; Tennyson was a frequent guest at Mirehouse, the Spedding family residence on Bassenthwaite, and spent his honeymoon at Tent Lodge near Coniston where he posed for a photograph by Lewis Carroll.

In 1846 Harriet Martineau built The Knoll in Ambleside; it was her home for the next 30 years, a power-house for her energetic support of good causes, her social reforms, her novels and children's books, her journalism, her lectures to the locals. Emerson called on her; John Bright was once caught on his knees measuring her study and sitting-room for a surprise present of carpets; Charlotte Brontë stayed for a week and left this tribute to a remarkable woman:

'She appears exhaustless in strength and spirits ... she is both hard and warm hearted, abrupt and affectionate, liberal and despotic ... she seems to be the benefactress of Ambleside, yet takes no sort of credit to herself for her active and indefatigable philanthropy.'

In 1871 John Ruskin bought Brantwood on the shore of Coniston, not even troubling to look at the house, convinced that 'any place opposite Coniston Old Man *must* be beautiful'. He found the building 'a mere shed of rotten timbers and loose stone' but set about transforming it into the modest mansion that it is today (Plate XIII). For Ruskin, it was a haven during the last 30 years of his life, years busy with various literary and artistic projects, years scarred by depression and mental illness—a sad end for a man described by Tolstoy as 'one of the most remarkable men, not only of England and our time but of all countries and all times. He was one of those rare men who think with their hearts'.

Canon H. D. Rawnsley also thought with his heart—and then took action. One of the founders of the National Trust, he was a fierce preservationist: he once led 2,000 protesters up Latrigg to establish public rights over a disputed footpath. He wrote prolifically about the Lake District and his *Literary Associations of the English Lakes* (1894) still remains an informative guide on the subject.

This century has witnessed a deluge of books with Lakeland associations. Some like O. S. MacDonnell's *George Ashbury*, Eliza Lynn Linton's *Lizzie Lorton of Greyrigg* and Edna Lyall's *Hope the Hermit* have retained their appeal even though their style grates on modern ears. Sir Hugh Walpole's 'Herries' saga has tempted many readers to explore its Lakeland setting; whatever Walpole's shortcomings as a writer, he genuinely loved the Lakes. Graham Sutton's five 'Fleming' novels have an authenticity that is rooted in their author's local links: born a Cumbrian, he displayed in his life and in his literature a deep affection for the fell country, buttressed by an encyclopaedic knowledge of its past. Melvyn Bragg, today's outstanding Cumbrian novelist, was born in Wigton; his fiction, like Graham Sutton's, reveals a sympathetic understanding of the Cumbrian character and the Cumbrian landscape, although he prefers the gentler scenery of fringe Lakeland for background.

Writers like Harry Griffin, Molly Lefebure and Jessica Lofthouse analyse Lakeland from different angles and with considerable success, but without doubt Lakeland's most eminent contemporary author is Norman Nicholson. He has lived in Millom on the edge of the National Park all his life, drawing his inspiration and material from the industrial belt of west Cumberland as well as from the glories of the fells. Primarily he is a poet, but the qualities animating his four volumes of verse—*Five Rivers, Rock Face, The Pot Geranium* and *A Local Habitation*—are apparent in his plays and topographical books. Indeed, no better introduction to the district, its history, its literature could be suggested than a study of Nicholson's *Cumberland and Westmorland, Portrait of the Lakes, The Lakers, Provincial Pleasures* and *Greater Lakeland*. Appropriately he contributes the opening chapter to this guide.

Many writers of children's books have turned to Lakeland as a setting for their stories—Arthur Ransome's *Swallows and Amazons*, Geoffrey Trease's Bannermere books, Marjorie Lloyd's *Fell Farm Holiday* and its sequels, Rosemary Sutcliff's *The Shield Ring*—to name only a handful. Beatrix Potter lived for many years at Sawrey, using the village and its immediate neighbourhood for her delightful animal fantasies. She achieved local distinction as a breeder of Herdwick sheep, and when she died she left a legacy of land and buildings to the National Trust.

It has been estimated that there are over 50,000 books that have some connexion with Lakeland. The number grows at the rate of nearly 100 a year: an alarming thought but at least a reassurance that the National Park may be enjoyed not only on the ground but also in print. For in this mountain of words lie the history and beauties of place, the aspirations and struggles of people, the variety and fascination of nature, the challenge and satisfaction of recreation.

14
Weather and Climate of the Lake Counties

by GORDON MANLEY

IN breezy temperate lands with the sea not far distant, the changing lights and sounds, the smells borne on the wind, and the feel of the air on our skins continually thrill or blunt our several senses; and in the Lake District throughout the year the mobile air among the mountains subtly adds to the pleasures of the visitor. The smell of the wet fellside mat grass on a gentle February morning; beyond the dark fells, the undying pink along the northern horizon on a clear midnight in June; the remorseless exhilarating grandeur of those November curtains of rain borne across Sty Head, or of the roaring south-wester; every year brings those Whitsuntide glories beloved of all North Countrymen, the onset of the colouring of the bracken in September, the 'breathless quiet of an October afternoon' that Dorothy Wordsworth enjoyed, or the 'loud autumn waters' when she went out after rain, the superb brilliance when the crisp bite of the Arctic air comes down in March.

On climate, as far as the visitor is concerned, Wordsworth himself in his *Description of the Scenery* (1822) said most of what was needed. He was well travelled, and wrote:

> 'I do not indeed know any tract of country in which, in so narrow a compass, may be found an equal variety in the influences of light and shadow upon the substance and beautiful features of the landscape . . . it has been ascertained that twice as much rain falls here as in many parts of the island, but the number of black drizzling days that blot out the face of things is by no means proportionately greater . . . nor is a continuance of thick flagging days so common as in the west of England or Ireland. The rain here comes down heartily and is frequently succeeded by clear bright weather . . . days of unsettled weather with partial showers are very frequent, but the showers darkening or brightening as they fly from hill to hill are not less grateful to the eye than finely interwoven passages of gay and sad music are touching to the ear.'

And he particularly commended late May to mid- or late June, and favoured late September and early October for the colour: a verdict which for many stands today.

For more than two centuries science not less than art and poetry has flowered in these counties in the minds of thoughtful men. Since 1756 temperature and rainfall have been recorded in one place or another. The Reverend Thomas Robinson, rector of Ousby, who wrote on the botany and the geology, was commenting on the helm wind in 1696. In 1786 Peter Crosthwaite at Keswick assisted John Dalton of Kendal in pioneer measurements of the height of the aurora; and Dalton went on to demonstrate his interest in the height of the clouds, the humidity of the air on Helvellyn, the persistence of the mountain snow. In 1823 blind John Gough of Kendal produced an original and thoughtful paper on the marked prevalence of dry north-easterly winds in spring.

Rainfall. Our theories of rainfall owe much to those early measurements around Lancaster and Kendal that were begun in 1784 by Campbell, in 1787 by Gough, and continued by Dalton, Harrison and Marshall. Gradually they were extended into the mountains after the day in 1844 when John Fletcher Miller from Whitehaven set up the first Seathwaite gauge, to surprise the London scientists with the size of its first year's catch. Seathwaite at the head of Borrowdale soon began to figure in the Victorian schoolbooks as the wettest place in England, and an exaggerated reputation of the Lake District for rain has long stood; but it needs a great deal of qualification. How many are aware that in February 1932 the very rainy head of Borrowdale recorded a month without measurable precipitation of any kind? And we now know that in a small area round Snowdon, and a larger area round Glengarry in the western Scottish highlands, greater annual totals are likely to occur.

Expansion of the air as it is forced upwards against the mountains leads to that cooling which produces cloud and, carried further, to that persistent heavy rainfall over many hours. On such a day ahead of a depression the coastal lowland in Furness or beside the inner Solway may have but a light shower or two. Moreover, as was first pointed out in 1852, the valleys radiating from the central knot give rise to further convergence and ascent of the airstreams from between south and west; hence there are about a dozen square miles between the heads of Eskdale and Borrowdale, Langdale and Wasdale, with an average exceeding 150 inches, reaching 185 locally beneath Great End. From beside the coast where Morecambe averages 38 inches and Barrow barely 40, the rainfall increases steadily to over 50 inches at Kendal, 60 by Windermere, 75 at Ambleside, over

Weather and Climate of the Lake Counties

90 at Grasmere. But as the air descends again beyond the mountains the decrease is rapid. In 6 miles, coming down Borrowdale, we find 131 inches at Seathwaite, just over 100 at Rosthwaite, 90 at Grange, 58 at Keswick. To the north-eastward the decrease is sharper: Penrith and Appleby average 35 inches, and along Edenside there are places with little more than 30. (For millimetres, multiply these figures by 25.) Effects are readily observed; going down Ullswater, the arable fields begin near Pooley Bridge and, in August, the ripening corn and the root crops will be seen.

Convergence of the airstreams from the south gives rise to small areas with over 100 inches on Fairfield—Helvellyn, and around the head of the river Kent into Mardale. From the sunny and relatively dry coastal strip by Seascale the rapid increase of rainfall into Ennerdale and Wasdale is conspicuous, and also the increasing dryness along the coast beyond St. Bees Head and Whitehaven.

But long ago it was perceived that most of the Lake District has about as much effective dry weather as anywhere else. Especially in spring and early summer the likelihood of anticyclonic weather with light N. to E. winds is considerable; then the Lake District lies in the shelter of the Scottish uplands and the Pennines. Skies are often remarkably cloudless, and in the clear air and strong sunshine south-facing slopes quickly dry out and burn, as the Ambleside gardeners know. Drought is far from uncommon.

Statistics of the number of days with 'measurable rain' (roughly, a 10-minute light shower) have long been available. Broadly, London has nearly 1 day in 2, Edinburgh 1 in 2, Manchester scarcely more. Bristol, Cardiff and Glasgow rank higher; but even the wetter parts of the Lake District run to little beyond 3 out of 5, Clearly those days that are wet, or showery, tend to be more so among the mountains. In the very wet centre there have been several occasions in the past century when around 8 inches has fallen in a day; but thunderstorms in southern England have been known to produce as much. It may be added that thunder is less frequently heard in the Lake District than in the East Midlands. March to June inclusive are the drier months, each with 5 to 6 per cent of the year's total; October to January have each 11 or 12 per cent, with January wettest in recent decades. With about 9 per cent August also ranks as rather wet, although it feels drier because evaporation is more rapid in the warmer air. As a rule April and June have fewest rainy days; December has most.

Snowfall. Measurable precipitation includes snow, always a matter of interest to visitor and resident alike. At Keswick or Ambleside the frequency of days with snow or sleet observed to fall is no greater than in the Manchester suburbs and indeed scarcely higher than in Croydon. Snow or sleet falls rather less frequently at the same level in the Lake District than on the Yorkshire Pennines, and is generally less in quantity; that is because it commonly falls on an easterly wind. For example, at 600 feet near Penrith the annual average is about 22 days, compared with 26 above Huddersfield and 31 at Ushaw near Durham. Coastal showers play their part nearer the North Sea.

The average annual number of days with snow-cover, that is, more than half the ground at the observer's level covered with snow at 9 a.m., is less than 5 on the coast, as the happy retired residents at Grange-over-Sands and Silverdale know. It rises to between 10 and 15 in the inland valleys, where chances are highest in January and February, comparable with Hampstead, Croydon, the lowland by Edinburgh, or Manchester airport. But with altitude the increase in both frequency and persistence of cover is rapid. Around the summit of the M6 motorway there are likely to be about 38 days; beside the Kirkstone Inn or the old Shap summit at 1,500 feet, about 50 days; and at 3,000 feet around Scafell Pike and Helvellyn, perhaps 110 days. These figures are derived from observations on the Northern Pennines, where the broad windswept summit of Cross Fell (2,930 feet) averages 106 and, 3 miles farther east, Moor House at 1,830 feet averages 72 between October and May; there, perhaps once in ten years there may be a short-lived cover in September or June. It is on record that Skiddaw was covered on the morning of 11 July 1888.

The range of variation is from about 1 morning only to about 60 in the valleys, and from 50 to 170 on the summit of Cross Fell. Because most notable falls come when the wind has an easterly component, falls are heavier and more lasting, not only on the Pennines, but in high-lying Matterdale facing NE, when compared with places like Wasdale or Dunnerdale. The chances of a spell of persistent snow-cover and hence some skiing above 2,000 feet are much higher on the Helvellyn Dodds than around Wasdale, and a ski-tow is sometimes operated at weekends by the Keswick enthusiasts on Great Dodd, towards which cars can be taken to over 1,300 feet. Skiing on the more snowy Pennines can be enjoyed sometimes into April, but the chances of low cloud and the raw wind forming an icy crust, or drifting the snow irregularly so that rocks and heather may be exposed, should not be overlooked. On most winter days the windy mountain climate, oscillating around the freezing point, is bitter indeed, and climbers need to take care.

Drifts accumulating in well-shaded northerly gullies linger, long after the summits have cleared; as a rule the last remnants can be found in places like Custs Gully on Great End, or at the top of the east-facing crags on Helvellyn and High Street. In 1951, after a cool spring marked by exceptional lack of warm rain, small drifts lasted on Helvellyn and Cross Fell until mid-July; it is possible that in 1879 some lasted well into August. But there is little support for the romantic ideas of some past tourists that snow could be found throughout the year.

Exceptional snowfalls in respect of depth and drifting occur less often in the Lake District than in Yorkshire or Durham. Perhaps the most renowned drifting fall on a strong south-easter was that in January 1940, after which no trains reached Barrow for four days. Celebrated falls in 1823 and 1767, and the blockage of the Furness Railway in March 1947 made their impression.

Drainage and evaporation. Many of the mountain slopes now turf-covered are underlain by the shattered screes of the Ice Age, hence they drain freely; yet within a few yards a rock-hollow will harbour a peat bog. Evaporation loss through the year resembles that of the adjacent counties, and the fact that at Keswick the early afternoon relative humidity in June is comparable with that at Norwich and less than that over much of Wales is a reminder that winds off the land are dry and, as Wordsworth said, 'thick flagging days' are less frequent than in parts of the West Country.

Temperature. Our mobile air ensures that averages of temperature through the winter months are much the same as elsewhere. In the familiar Fahrenheit degrees, January stands close to 39° at Keswick as it does at London, Edinburgh or Manchester. But in common with the rest of northern England, summer is a little cooler and fresher than in the Midlands and the Thames Valley; and among the mountains it is inclined to be more cloudy. July's average overall at Keswick stands just on 60°, compared with Manchester above 61°, Edinburgh above 59°, and 64° in the London suburbs. But it is the lower afternoon maxima that reflect the fresher breeze and proximity to the sea that gladdens the heart of the vigorous holidaymaker escaping from the stale air of the stuffy city.

Throughout the year the fresh clean air is notable. Pollution is not absent, although local sources in Barrow and Whitehaven are small. It can come from Glasgow and Teesside, as well as the West Riding and south Lancashire, just as the London smoke reaches Norwich and Southampton. But in the Lake District the lichens, sensitive to polluted air, are still growing on the walls; and the many evergreens add to the variety of colours, especially in winter.

It is however a concomitant of our breezy maritime climate, with its frequent cloud and stronger wind over the higher summits, that with altitude daytime temperatures fall more rapidly than those at night; hence the rate of change with altitude in the effective warmth, as far as growth is concerned, is decidedly rapid. This explains why the tree-line at about 2,000 feet is so low compared with the Alps or even inland Norway. Hence the craggy summits at a mere 3,000 feet stand to our eyes as mountains. In sheltered gullies away from the nibbling sheep occasional birch and rowan can be found almost up to 2,000 feet, for example, above Codale Tarn or at the head of Grainy Gill.

Spring as a rule comes a little later than in the South Midlands, but in a mild season daffodils can flower in February, even as far inland as Windermere. Altitude and exposure are most effective in the spring months, very noticeably if the season is cold; it is then that the sharpest local contrasts occur between sunlit valley and icy mountain gully, as the Easter climbers know. In high Matterdale, or at Alston farther east, the daffodils will linger until late May. Moreover there are acute local differences in the incidence and severity of frost. From the available observations comparisons are readily made of the average annual number of mornings when the minimum air temperature in the screen has fallen below freezing-point. In open country, *e.g.* on Midland airfields, about 65 can be expected (Cambridge 63). At Keswick on the open ground between the lakes, the average is 55; at Ambleside, 62; at Urswick-in-Furness, 56. With the first sharp frost coming towards the second week in October, there will probably be four in April, and the last should come about the second week in May.

But in any mountain country on a still clear evening the downward trickle of cool air is perceptible, as is the manner in which it 'ponds' on the valley-floor, or among the many irregularities that are particularly conspicuous along Edenside. The whole district abounds in frost-hollows; watch the behaviour of the sheep, who not only like to nibble the better-drained roadside grass, and to enjoy the warmth of the road after a day's sunshine; on a quiet night, they settle on the available rises and the hummocks.

In Grizedale the Forestry Commission records temperatures in a screen at the foot of long slopes close to the valley-floor. Over the same period it has averaged 90 nights with frost, compared with 62 at Ambleside 6 miles distant. Near Penrith the average is 72, but from the older Appleby station in a garden below the castle, the record indicates about 84. So Appleby may well get a June nip that Ambleside will miss. Out along the coast Morecambe averages 38, Sellafield 42; hence it is probable that in gardens on the favoured slopes at Grange there may well be as few as 30, yet only 2 miles away in the flat fields by Cartmel there are over 60. Down on the open mosses by Gilpin Bridge frost is sharper and more frequent by contrast with the famous old walled and cultivated garden at Levens.

Extremes can occur, in the brilliant dry anticyclonic weather so characteristic of early June, or later when the hot dry south-easter comes up from the Continent. Twice in this century 90° has just been surpassed at Keswick. By contrast, given quiet clear winter nights with a deep snow-cover, the mercury has occasionally fallen below zero (F.) in the valleys. In January 1940, when the minimum at Keswick was zero, at Ambleside −6° was recorded; and towards the Eden, reports from local observers support the likelihood that −10° may not be unknown. But what is much more serious for the gardener is the varying incidence of frost in spring, already mentioned. Choice of site and care in selection of seed and time of sowing has long been known to pay. To have a garden on the slopes adjacent to the deeper lakes can be an advantage, except when the rather rare winter arrives, as in 1963, when even the larger lakes such as Windermere became completely frozen.

Sunshine, wind, cloud and fog. Although rainfall totals may appear high, the proportion of clear and sunny weather is little affected. Out on the coast the recorded average annual duration is good, exceeding 1,500 hours yearly; remember also that if the sun is very close to the horizon it will not be registered. Generally in Britain, by reason of greater cloud amount, inland stations record about 10 per cent less than the nearest sea-coast, and among the mountains some low-angle sunshine, especially in winter, is directly cut off. Hence the annual average of 1,232 hours at Keswick, rising towards 1,350 at more open Penrith and to about 1,400 at Lancaster and Carlisle, is between 30 and 35 per cent of 'possible' and compares well with expectation for the latitude. The higher mountain summits are much more cloudy, with scarcely two-thirds of the sunshine commonly registered in the lowlands nearby.

May and June rank highest with nearly 40 per cent; hence with the long light, June averages about seven hours daily. In common with the rest of Britain, July and August are more cloudy, September a little less as a rule. December with its short days is also stormy and cloudy, and little more than an hour of sunshine a day can be expected. Let it not be forgotten, however, that even August can occasionally do remarkable things; in 1947 nearly 70 per cent of the possible, over 10 hours a day, was recorded at Keswick, with but one brief shower; and on the Cumberland coast the wonderful sunny cold February of 1963 was equally remarkable.

At low levels wind is much less strong inland than at sea or on the coast, but it is often gusty and local variations can be conspicuous, for example, when a south wind blows along the length of Windermere. On the more sheltered west side the trees retain their leaves well into the autumn and colouring can be very good. Acceleration over the rounded summits and through the high passes ('Windy Gap' for example) is evident. A brief record from the bare rounded summit of Dun Fell indicates an average wind speed of force 5 (21 mph), about double that characteristic of the open lowlands. A mean hourly wind speed of 99 mph and a gust of 134 mph have been measured. Everywhere on the tops winter gales can be severe.

Local developments are noteworthy. The 'helm wind' of Cumberland and Westmorland is often mentioned; this is a localised acceleration of winds from between NNE and E down the great scarp formed by the Northern Pennines, most notably around Cross Fell. The phenomenon is broadly analogous to the behaviour of a stream of water flowing over a submerged weir; the air accelerates down the escarpment (the 'weir') and forms a 'standing wave', often marked by a bar of cloud at its crest, parallel to the cloud (the 'helm') which characteristically stands above the main Pennine ridge, and distant 3 to 7 miles from it, the intervening space being clear. Under a characteristic overcast sky with a brisk north-easter blowing force 4–5 at Tynemouth, the airflow down the escarpment can attain gale force (8), descending as a steady roaring torrent. Near the foot of the fellside it rises again, and a mile farther west under the 'bar' there is often nearly a calm. The rising stream can be used by gliders. With a cloudy north-easter, downslope acceleration, of the gusts on the lee side of hills tends to occur locally farther west, where the term 'helm wind' is sometimes used, but the more irregular terrain means that it does not develop the same strength and steadiness as it does off the Pennines. It can be reconnoitred with a car on favourable days, most commonly from March to May. The dry, cutting north-easter is not generally liked, either by the farmer, his wife, or his sheep.

Above Honister Pass the quarrymen refer to 'the wind in the crack'; it appears that, given a strong wind from a particular direction, local suction may suddenly tear the men from their ledges. That perhaps resembles the outward explosion of windows on the lee side of gale-swept tower blocks. And with a light wind after fine weather the appearance and growth or otherwise of the 'Borrowdale sop'—the first onset of cloud over Sty Head, seen to the southward—is watched as a possible precursor of rain.

In windy damp weather low scurrying cloud mantling all the hills is characteristic, but on the upland roads the hindrance to traffic is slight unless it develops into thick drizzle or snow. In the valleys, thick fog of the normal Midland winter variety is decidedly infrequent; not only is the air clean, but it is rarely stagnant.

Lake District weather adds to the pleasures for the thoughtful; its effects can be savoured; and testimony to the climate can indeed be sought in older writings on the vigour of the long-lived dalesmen and their traditionally comely daughters, in the lively intelligence, in the contentment that so many find who come to reside here in later life.

Some Climatological Statistics

KESWICK (254 ft.): *Averages of Temperature (1931–1958) and Rainfall (1916–1950)*

	Jan	Feb	Mar	Apr	May	June	July	Aug	Sept	Oct	Nov	Dec	Year
Monthly (°C)	3·7	3·8	5·8	7·8	10·9	13·7	15·2	15·0	12·8	9·8	6·7	5·0	9·2
Means: (°F)	38·7	38·8	42·4	46·2	51·6	56·7	59·4	59·0	55·0	49·6	44·1	41·0	48·6
Mean daily Max: (°C)	6·7	6·8	9·3	11·7	15·3	18·0	18·9	18·8	16·4	13·0	9·7	7·9	12·7
(°F)	44·0	44·2	48·7	53·1	59·5	64·4	66·0	65·9	61·5	55·4	49·5	46·2	54·9
Mean daily Min: (°C)	0·6	0·7	2·2	4·0	6·5	9·5	11·6	11·1	9·2	6·5	3·8	32·1	5·7
(°F)	33·1	33·3	36·0	39·2	43·7	49·1	52·8	51·9	48·5	43·7	38·8	5·8	42·3

Approximate average extremes for each month (°F):

	Jan	Feb	Mar	Apr	May	June	July	Aug	Sept	Oct	Nov	Dec	Year
	52	52	58	64	72	76	78	75	70	64	57	53	80
	21	22	26	29	33	40	44	43	37	31	27	23	16

Extremes on record for Keswick are (°F): 91° in July 1948; 0° in January 1940. Earlier records give 92° at Kendal (1868 and 1921): −6° at Ambleside (1940); −8° near Carlisle (1892).

Rainfall at Keswick

(inches)	Jan	Feb	Mar	Apr	May	June	July	Aug	Sept	Oct	Nov	Dec	Year
	6·7	4·2	3·4	3·3	3·2	3·3	4·4	5·1	5·6	6·8	6·1	6·0	58·1

Upland and Mountain temperatures must be estimated from the Pennine stations to E. Over 27 years at Moor House (1,830 ft.) and 6 years at summit of Dun Fell (2,780 ft.) we have:

Monthly Means (°F):

	Jan	Feb	Mar	Apr	May	June	July	Aug	Sept	Oct	Nov	Dec	Year
Moor House	31·5	31·8	35·0	38·8	45·0	50·3	53·3	53·0	49·2	42·8	34·6	37·2	41·9
Dun Fell	28·0	28·0	31·3	35·0	40·5	46·3	49·5	49·0	45·5	39·5	34·5	31·0	38·2

The length of the season during which, with a temperature hovering frequently around freezing-point, raw damp cold with a penetrating wind is likely to be experienced should be noted.

RAINFALL

Average totals are given in the text, in inches: $1\ in.=25\cdot 4\ mm$. The average percentage of the annual total that falls in each month is much the same at wetter or drier stations and can here be summarised:

Jan	Feb	Mar	Apr	May	June	July	Aug	Sept	Oct	Nov	Dec
12	7	6	6	5½	5½	7	9	9	11	11	11

SUNSHINE DURATION

At Keswick:

	Jan	Feb	Mar	Apr	May	June	July	Aug	Sept	Oct	Nov	Dec	Year
hours	34	56	102	139	192	181	146	140	103	72	39	28	1232

Some is here cut off by surrounding hills. Newton Rigg by Penrith records nearly 10% more, Seascale on the coast about 25% more.

SNOW-COVER

Some average annual totals for snow or sleet falling and for snow-cover are given.
Valley-floors can be expected to be covered on about 12–15 mornings yearly.
Observations are available for the summit of Cross Fell (2,930 ft.) which over the past 22 years has been noted as covered on an average of 106 days. Scattered drifts last longer. For the months September–June, averages are: $0\cdot 1$, 2, 11, 20, 23, 21, 16, 9, 4, $0\cdot 1$ days.

SHORT BIBLIOGRAPHY

ABERCROMBIE, P., and KELLY. *Cumbrian Planning Guide.* London, 1932.
ABRAHAMS, Mrs ASHLEY P. *Poems of Lakeland.* London, 1946.
ARMSTRONG, A. M., MAWER, A., and DICKINS, BRUCE. *The Place-Names of Cumberland I-III* (English Place-Name Society, Vols. XX-XXII). Cambridge, 1945.
BADDELEY, M. J. B. *The English Lake District.* London, 1902.
BAKER, J. G. *The Flora of the English Lake District.* London, 1885.
BLEZARD, E. *Birds of Lakeland.* Arbroath, 1943.
BLEZARD, E. *Lakeland Natural History.* Arbroath, 1946.
BOUCH, C. M. L. *Prelates and Peoples of the Lake Counties.* Kendal, 1948.
BOUCH, C. M. L., and JONES, G. P. *The Lake Counties, 1500-1830.* Manchester, 1961.
BOUCH, C. M. L., and JONES, G. P. *A Short Economic and Social History of the Lake Counties.* Manchester, 1961.
COLLINGWOOD, W. G. *The Lake Counties.* London, 1949.
COLLINGWOOD, W. G. *Lake District History.* Kendal, 1928.
COOPER, W. HEATON. *The Hills of Lakeland.* London, 1938.
COOPER, W. HEATON. *Tarns of Lakeland.* London, 1960.
DAVIES-SHIEL, M., and MARSHALL, J. D. *Industrial Archaeology of the Lake Counties.* David & Charles, 1969.
DICKINSON, W. *Cumbriana.* London and Whitehaven, 1875.
EDEN, F. M. *The State of the Poor* (3 vols.). London, 1797.
EVANS, A. L. *The Naturalists' Lake District.* Dalesman Publications.
FELL, A. *The Early Iron Industry of Furness and District.* Ulverston, 1906.
FELL, C. *Early Settlement in the Lake Counties.* Silsden, 1970.
Fell and Rock Climbing Club. Rock Climbing Guides (second series) in 8 vols. 1948-52.
GARLICK, T. *Romans in the Lake Counties.* Silsden, 1970.
GARNETT, F. W. *Westmorland Agriculture, 1800-1900.* Kendal, 1912.
GRIFFIN, A. H. *Inside the Real Lakeland.* Preston, 1961.
GRIFFIN, A. H. *In Mountain Lakeland.* Preston, 1963.
GRIFFIN, A. H. *Pageant of Lakeland.* Hale, 1966.
GRIFFIN, A. H. *The Roof of England.* Hale, 1968.
GRIFFIN, A. H. *Still the Real Lakeland.* Hale, 1970.
HERVEY, G. A. K., and BARNES, J. A. G. *Natural History of the Lake District.* Warne, 1970.
HODGSON, H. W. *Bibliography of the History and Topography of Cumberland and Westmorland.* Joint Archives Committee for Cumberland and Westmorland and Carlisle, 1968.
HOUSMAN, J. *Topographical Description of Cumberland and Westmorland, etc.* Carlisle, 1800.
HUGHES, E. *North Country Life in the Eighteenth Century* (Vol. II). Oxford, 1965.
Institute of Geological Sciences. *British Regional Geology: Northern England* (4th edition). H.M.S.O., 1971.
JOY, D. *Cumbrian Coast Railways.* Clapham, 1968.
LAMB, H. H. *The English Climate.* English Universities Press, 1964.
LEFEBURE, M. *The English Lake District.* Batsford, 1964.

LEFEBURE, M. *Cumberland Heritage*. Gollancz, 1970.
MACAN, T. T., and WORTHINGTON, E. B. *Life in Lakes and Rivers*. London, 1951.
MACPHERSON, J. *Fauna of Lakeland*. Edinburgh, 1892.
MANLEY, G. *Climate and the British Scene* (5th impression), Collins New Naturalist; also available in Fontana (paperback). London, 1972.
MARR, J. E. *The Geology of the Lake District*. Cambridge, 1916.
MARSHALL, J. D. *Furness and the Industrial Revolution*. Barrow, 1958.
MARSHALL, J. D. *Old Lakeland*. David & Charles, 1971.
MILLWARD, R., and ROBINSON, A. *The Lake District*. Eyre and Spottiswoode, 1970.
MILLWARD, R., and ROBINSON, A. *Cumbria*. Macmillan, 1972.
MITCHELL, G. H. *Geological History of the Lake District*. Yorkshire Geological Society, 1956.
MONKHOUSE, P. J. *The English Lake District*. Sheffield, 1957.
NICHOLSON, N. *Cumberland and Westmorland*. Hale, 1949.
NICHOLSON, N. *Portrait of the Lakes*. Hale, 1965.
NICHOLSON, N. *The Lakers*. London, 1972.
PALMER, J. H. *Historic Farm Houses in and around Westmorland*. Kendal, 1952.
PEARSALL, W. H. *Mountains and Moorlands*. London, 1950.
PEARSALL, W. H., and PENNINGTON, W. *The Lake District—a Landscape History*. Glasgow, 1973.
PEVSNER, Sir N. *Cumberland and Westmorland*. Penguin, 1967.
ROLLINSON, W. *A History of Man in the Lake District*. London, 1967.
SHACKLETON, E. H. *Lakeland Geology*. Clapham, 1973.
SHAW, W. T. *Mining in the Lake Counties*. Clapham, 1970.
SMITH, A. H. *The Place-Names of Westmorland* (English Place-Name Society, Vols. XLII–XLIII). Cambridge, 1967.
SMITH, K. *Early Prints of the Lake District*. Nelson, 1973.
SYMONDS, H. H. *Walking in the Lake District*. Edinburgh, 1962.
THOMPSON, BRUCE L. *Prose of Lakeland*. London, 1954.
THOMPSON, BRUCE L. *The Lake District and the National Trust*. Kendal, 1946.
Wainwright's Pictorial Guides. *The Lakeland Fells* (7 vols.). Kendal, 1955–63.
WARD LOCK. *Guide to the English Lakes*. London, 1891.
WATSON, J. *The English Lake District Fisheries*. London, 1925.
WORDSWORTH, W. (ed. de Selincourt). *Guide to the Lakes*. London, 1930.
YAPP, W. B. *Birds and Woods*. Oxford, 1961.
YOUNG, A. *Six months' tour through the North of England* (Vol. III). London, 1771.

APPENDICES

I. Scheduled Ancient Monuments in the Lake District National Park

1in. Tourist Map
Grid Reference

BURIAL MOUNDS AND MEGALITHIC MONUMENTS

Askham, Cop Stone and standing stones, Askham Fell	494220/496216
Askham, round barrows on Askham Fell	493222/489224/488225
Askham and Barton, round barrows on Threepow Raise	481220/482219
Bampton, two cairns on Bampton Common	493164/501164
Bampton, standing stones on Four Stones Hill	491164
Barton, The Cockpit, stone circle	483222
Barton, stone circle on Swarth Fell	457193
Bootle, Bootle Fell, cairns 1,000 yards east of Low Kinmont	127896
Bootle, Bootle Fell, groups of cairns	132892
Bootle, Bootle Fell, Little Grassoms cairns	133890
Coniston, cairns on The Rigg, Banishead	284967
Coniston, cairns on Foul Scrow	292982
Ennerdale and Kinniside, group of cairns on Stockdale Moor	101082
Ennerdale and Kinniside, round barrows at Smithy Beck	120147
Ennerdale and Kinniside, Sampson's Bratful, oval mound	099080
Eskdale, stone circles on Burn Moor near Boot	174024
Eskdale and Muncaster, cairns on and ¼ mile south-west of Water Crag	154974/152969
Hawkshead, round barrow ¼ mile south-west of Thompson Ground	340983
Hutton, round barrow on Great Mell Fell	396255
Ireby, round barrow on Aughertree Fell	260380
Irton with Santon, cairn field in Mecklin Park	125020
Kirkby Ireleth, Giant's Grave, round barrow	256880
Kirkby Ireleth and Lands Common to Lowick and Subberthwaite, two ring cairns on Gawthwaite Moor	264856/264849
Kirkby Ireleth and Subberthwaite, cairns on Heathwaite Fell	255875
Millom, Sunkenkirk Circle, Swinside Fell	171882
Nether Wasdale, Grey Borran, group of cairns	122065
Nether Wasdale, Yokerill Haws, groups of seven cairns	114075

1in. Tourist Map
Grid Reference

St. John's Castlerigg and Wythburn, stone circle on Castlerigg	292236
St. John's Castlerigg and Wythburn, cairn on Dunmail Raise	322118
Setmurthy, stone circle on Elva Hill	179318
Torver, dike circles and cairns on Bleaberry Haws	267947
Torver, cairns on Throng Moss 700 feet south-west of reservoir	278922
Troutbeck, two cairns 700 feet south of Bluegill Fold	425077
Troutbeck, stone circle ¾ mile north-west of Troutbeck Park	420075
Waberthwaite, Corney Fell, group of cairns 400 yards east of Buckbarrow Bridge	133929
Waberthwaite, Corney Fell, group of cairns 400 yards south of Charlesground Gill	130904
Waberthwaite, cairns south-east of High Kinmont	126906
Waberthwaite and Muncaster, stone circle and cairns north of Whiterow Beck	133941

CAMPS AND SETTLEMENTS

Askham, settlement at Skirsgill Hill	499232
Bampton, settlement on Four Stones Hill	490163
Bampton, enclosure on Knipescar Common	529195
Bampton, Castle Crag, Mardale	469128
Bampton, earthwork 220 yards north-east of Measand Bridge	490156
Bampton, earthwork in Scarside Plantation	530193
Borrowdale, Castle Crag	251159
Borrowdale, Reecastle Crag settlement	277176
Broughton West, settlement on The Hawk	241923
Caldbeck and Mungrisdale, Carrock Fell hill-fort	342337
Coniston, enclosure on The Rigg, Banishead	284967
Coniston, settlement on Greenbury Beck 400 yards south of Fell Foot Farm	300029
Dacre, Dumallet hill-fort	468246
Ennerdale and Kinniside, settlements on Tongue Haw	074098
Hugill, settlement near High Borrans	438010
Ireby, camp south-east of Whitefield House	248345
Ireby, settlements on Aughtertree Fell	260380
Kentmere, Millrigg settlement	462025
Lowther, settlement in Cragside Wood, Lowther Park	525216
Martindale, settlement on Heck Beck	423154
Muncaster, Barnscar British settlement	136960
St. John's Castlerigg, Shoulthwaite Gill hill-fort	300188
Shap Rural, settlement 400 yards south-west of Naddle Bridge	507157
Troutbeck, settlements north and south of Sad Gill	423085/423081
Windermere, settlement on Allen Knott	415011
Wythop, Castle Haw	201308

Appendix I: Scheduled Ancient Monuments

1in. Tourist Map Grid Reference

ROMAN REMAINS

Ambleside, Roman fort	372034
Bewaldeth and Snittlegarth, Caermote Roman forts	202368
Eskdale, Hardknott Castle, near Brotherilkeld	218015
Eskdale, Wrynose and Hardknott Passes	235014
Hutton, small Roman fort ⅓ mile north-east of Troutbeck station	382273
Muncaster, Ravenglass Roman fort	088959
Muncaster, Roman kilns	131986
Muncaster, Walls Castle, Ravenglass	088960
Mungrisdale, Roman temporary camp south of Fieldhead Farm	379273

ECCLESIASTICAL BUILDINGS

Bootle, Seaton Nunnery (site of)	106898
Crook, tower of ruined church of St. Catherine	449946
Over Staveley, tower of ruined chapel of St. Margaret	473981
St. Bridget Beckermet, Calder Abbey	051064
St. Bridget, Beckermet, Calder Abbey (additional area)	050064
Shap Rural, Shap Abbey	548152

CROSSES AND INSCRIBED STONES

Gosforth, Gosforth cross	072036
Irton and Santon, Irton churchyard cross	091004
Waberthwaite, cross-shafts in churchyard	100952

CASTLES AND FORTIFICATIONS

Dacre, Dacre Castle, earthworks of	461266
Kirkby Ireleth and Subberthwaite, settlement on Heathwaite Fell	255875

INDUSTRIAL MONUMENTS, INCLUDING WINDMILLS AND WATERMILLS

Colton, Stott Park bobbin mill, Finsthwaite	373883
Coniston, mill and dwelling on river Brathay 600 feet south-west of Fell Foot Farm	298031
Coniston, bloomeries in Water Park	302955/303956
Millom Without, Duddon furnace	197883
Upper Allithwaite, Wilkinson's Monument, Lindale	419802

OTHER SECULAR SITES AND BUILDINGS

Askham, earthwork east of Setterah Park	514212
Bampton, Towtop Kirk	493179
Hawkshead, Hall gatehouse	349988
Keswick Town Hall	263237
Shap Rural, deerpound near Tonguerigg Gill	536156

1in. Tourist Map Grid Reference

BRIDGES

Askham and Bampton, bridge over Heltondale Beck 550 feet south of Widewath	502208
Askham and Bampton, bridge over Heltondale Beck 250 yards south-south-west of Widewath	501208
Bampton, stone bridge, Cawdale Beck, west of Bampton	496178
Colton, Newby Bridge	369863
Drigg and Carleton, Drigg Holme packhorse bridge	077988
Ennerdale and Kinniside, Monks Bridge	063103

DESERTED SETTLEMENT

Barton, moated site south of Gale Bay	465234

II. Some Useful Addresses

NOTE: The addresses contained in this Appendix are likely to change from time to time. For the most recent revisions to this list, please apply to either the Countryside Commission or the Lake District National Park Information Service.

ANGLING

Cumberland River Authority
256 London Road, Carlisle. *Tel.:* Carlisle 25151/2.
Lancashire River Authority:
Fishery Department, Mill Lane, Halton-on-Lune, near Lancaster. *Tel.:* Halton-on-Lune 473/4.

ARCHAEOLOGICAL, GEOLOGICAL, AND NATURAL HISTORY SOCIETIES

Cumberland and Westmorland Archaeological Society
Hon. Sec.: W. Rawlinson, Institute of Extension Studies, University of Liverpool, P.O. 147, 1 Abercrombie Square, Liverpool LV69 3BX.
Cumberland Geological Society
Hon. Sec.: R. E. O. Pearson, 31 Main Street, St. Bees.
Tel.: St. Bees 461.
Lake District Naturalists' Trust
Hon. Sec.: Wing Cdr. J. G. Lingard, 5 Annisgarth Close, Windermere.
Tel.: Windermere 4219.
Ambleside Field Society
Hon. Sec.: A. Brooks, 3 St. Mary's Park, Windermere.
Tel.: Windermere 2896.
Arnside and District Natural History Society
Hon. Sec.: Miss G. B. Dale, The Cottage, Uplands, Arnside.
Tel.: Arnside 761295.
Barrow Field Naturalists' Club
Hon. Sec.: J. Kellett, 14 Monksvale Grove, Barrow-in-Furness.
Carlisle Natural History Society
Hon. Sec.: G. Horne, 17 Yetlands, Dalston, Carlisle.
Tel.: Dalston 710081.
Eden Field Club

Hon. Sec.: E. Hinchcliffe, Mill Cottage, Murton, Appleby.
Tel.: Appleby 51430.

Grange-over-Sands Natural History Society
Hon. Sec.: Dr F. H. Day, Barn Top, Lindale, Grange-over-Sands.
Tel.: Grange 3295.

Kendal Natural History Society
Hon. Sec.: Miss K. Littlewood, 6 Rowan Tree Crescent, Kendal.
Tel.: Kendal 20779.

West Cumberland Field Society
Hon. Sec.: A. B. Warburton, 2 Monk Moors, Eskmeals, Bootle Station.

Penrith Natural History Society
Hon. Sec.: C. Denwood, 33 Carleton Drive, Penrith.

Coniston Natural History Society
Hon. Sec.: R. B. Cluett, 12 Collingwood Close, Coniston.

Keswick Natural History Society
Hon. Sec.: G. Wilson, Broughton Grove, Rogerfield, Keswick.
Tel.: Keswick 73153.

AUTOMOBILE ASSOCIATION

Headquarters:
Fanum House, Basing View, Basingstoke, Hants.
Tel.: Basingstoke 20123.

Carlisle:
Townpoint, 37 Castle Street, Carlisle.
Tel.: Carlisle 24274.

Patrol Points:
Carnforth 2036.
Penrith 3217.

24-hour Service:
Manchester—*Tel.:* (061) 485-6299.
Newcastle—*Tel.:* Newcastle 28611.

CAMPING CLUB OF GREAT BRITAIN AND IRELAND

Headquarters:
11 Lower Grosvenor Place, London SW1
Tel.: (01) 828-9232.

CARAVAN CLUB OF GREAT BRITAIN AND IRELAND

Headquarters:
65 South Molton Street, London W1Y 2AB.
Tel.: (01) 629-6441.

Appendix II: Some Useful Addresses 147

Cumberland and Westmorland Group:
  Hon. Sec.: D. Graham, 460 London Road, Carlisle.
  Tel.: Carlisle 26345 (home), 26288 (office).

CENTRAL COUNCIL OF PHYSICAL RECREATION

Headquarters:
  70 Brompton Road, London SW3 1HE.
  Tel.: (01) 584-6651/2.
North-Western Regional Office:
  Ralli Building, Stanley Street, Salford M35 FJ.
  Tel.: (061) 834-0338/9573.

COUNCIL FOR NATURE

Headquarters:
  Zoological Gardens, Regent's Park, London NW1.
  Tel.: (01) 722-7111.

COUNCIL FOR THE PROTECTION OF RURAL ENGLAND

Headquarters:
  4 Hobart Place, London SW1W 0HY.
  Tel.: (01) 235-9481.
Lancashire Branch:
  Secretary: Samlesbury Hall, Preston New Road, Samlesbury PR5 0UP.
  Tel.: Mellor 2010.
Craven Branch:
  Hon. Sec.: D. Joy, Hole Bottom, Hebden, Skipton, North Yorkshire.
  Tel.: Grassington 369.

COUNTRYSIDE COMMISSION

Headquarters:
  John Dower House, Crescent Place, Cheltenham, Glos. GL50 3RA.
  Tel.: 0242 21381.

FIELD STUDY CENTRES

Brathay Field Study Centre, Old Brathay, Ambleside.
  Tel.: Ambleside 3041.
Y.H.A., Esthwaite Lodge, Hawkshead, Ambleside.
  Tel.: Hawkshead 293.
Y.H.A., High Close, Loughrigg, Ambleside.
  Tel.: Langdale 212.

148   *Lake District*

High Borrans Field Study Centre (Tynemouth Education Authority), Windermere.
Tel.: Windermere 2816.
Brantwood Trust, Brantwood, Coniston.
Tel.: Coniston 396.

FORESTRY COMMISSION

Headquarters:
25 Savile Row, London W1X 2AY.
Tel.: (01) 734-0221.
North-West Conservancy:
North Lakes District Office, 4 Main Street, Cockermouth.
Tel.: Cockermouth 2063.
South Lakes District Office, 35 Stricklandgate, Kendal.
Tel.: Kendal 22587.
Grizedale Forest Office (*Visitor Facilities*):
Satterthwaite, Hawkshead, near Ambleside.
Tel.: Satterthwaite 214.

FRIENDS OF THE LAKE DISTRICT

Secretary:
G. V. Berry, Kilns, Greenside, Kendal.
Tel.: Kendal 20296.

GLIDING

British Gliding Association: Artillery Mansions, 75 Victoria Street, London SW1.
Tel.: (01) 799-7548.
Lakes Gliding Club:
Hon. Sec.: R. C. Bull, 13 Saunders Close, Walney, Barrow-in-Furness.
Tel.: Barrow 41298.

GOLF

Windermere Golf Club:
Cleabarrow, Windermere.   *Tel.:* Windermere 3123.
Cockermouth Golf Club:
Embleton, Cockermouth.   *Tel.:* Bassenthwaite Lake 223.

HOLIDAY FELLOWSHIP

Headquarters:
142 Great North Way, Hendon, London NW4.
Tel.: (01) 203-3381.

Appendix II: Some Useful Addresses

Lake District Area:
  Hon. Sec.: Mrs H. M. Phillipson, 45 Aynam Road, Kendal.
Hostels:
  Derwent Bank, Portinscale, Keswick. *Tel.:* Keswick 72326.
  Monk Coniston Hall, Coniston. *Tel.:* Coniston 342.
  Newlands, Stair, Keswick. *Tel.:* Braithwaite 220.

LAKE DISTRICT PLANNING BOARD

Headquarters:
  County Hall, Kendal. *Tel.:* Kendal 21000.
Head Warden and Information Service:
  Bank House, High Street, Windermere. *Tel.:* Windermere 2498.

MOUNTAIN RESCUE

Lake District Mountain Accidents Association:
  Hon. Sec.: B. Eales, 6 Meadowfield, Gosforth.

NATIONAL TRUST

Headquarters:
  42 Queen Anne's Gate, London SW1. *Tel.:* (01) 930-0211.
Northern Area Office:
  Broadlands, Borrans Road, Ambleside. *Tel.:* Ambleside 3003/4.

NATURE CONSERVANCY COUNCIL

Headquarters:
  19 Belgrave Square London SW1X 8PY. *Tel.:* (01) 235-3241.
North Regional Office:
  Merlewood Research Station, Windermere Road, Grange-over-Sands. *Tel.:* Grange 2264/5.

NATURE RESERVES

Drigg Dunes, near Ravenglass:
  County Land Agent, Cumbria County Council, 1 Alfred Street North, Carlisle CA1 1PX.
  Tel.: Carlisle 23456 (Ex. 244).
Forestry Commission:
  Grizedale Forest Office, Satterthwaite, Hawkshead, near Ambleside.
  Tel.: Satterthwaite 214.
Lake District Naturalists' Trust:
  Hon. Sec.: Wing Cdr. J. G. Lingard, 5 Annisgarth Close, Windermere.
  Tel.: Windermere 4219.

150   *Lake District*

OUTDOOR PURSUITS CENTRES

Brathay Field Study Centre:
   Old Brathay, Ambleside.   *Tel.:* Ambleside 3041.
Denton House Outdoor Pursuits Centre (Cumbria Education Authority)*:*
   Penrith Road, Keswick.   *Tel.:* Keswick 72137.
Derwent Hill Outdoor Education Centre (County Borough of Sunderland Education Committee)*:*
   Portinscale, Keswick.   *Tel.:* Keswick 72005.
Eskdale Outward Bound Mountain School:
   Eskdale Green.
   Tel.: Eskdale 281; Warden's House 282.
Ghyll Head Outdoor Pursuits Centre (Manchester Corporation)*:*
   Ghyll Head, Windermere.
   Tel.: Windermere 3751 and 2696.
Hawse End Outdoor Pursuits Centre (Cumbria Education Authority)*:* Authority)*:*
   Portinscale, Keswick.
   Tel.: Keswick 72816.
Howtown Outdoor Activities Centre (Durham Education Authority)*:*
   Ullswater, Penrith.
   Tel.: Martindale 208.
Outward Bound Mountain School (Ullswater) Ltd.:
   Pooley Bridge, Penrith.
   Tel.: Pooley Bridge 347.
Tower Wood Outdoor Pursuits Centre (Lancashire County Education Committee)*:*
   Tower Wood, Windermere.
   Tel.: Newby Bridge 519.
Y.M.C.A. National Training Centre:
   Lakeside, Newby Bridge, Ulverston.
   Tel.: Newby Bridge 352.

PRINCIPAL TRANSPORT UNDERTAKINGS

Ribble Motor Services Ltd.:
   Frenchwood, Preston PR1 4LU.   *Tel.:* Preston 54754.
         Enquiry Offices: Ambleside—*Tel.:* 3233.
                           Kendal   —*Tel.:* 20932.
                           Keswick —*Tel.:* 72791.
Cumberland Motor Services Ltd.:
   Tangier Street, Whitehaven.   *Tel.:* Whitehaven 3781.
         Enquiry Office: Keswick—*Tel.:* 72791.

Appendix II: Some Useful Addresses 151

Mountain Goat Minibus Service:
  54 Main Road, Windermere.
  Tel.: Windermere 4106; evenings and weekends 4527.
Keswick Minibus and Taxi Service:
  23 Helvellyn Street, Keswick. *Tel.:* Keswick 72676.
British Rail:
  Ladywell House, Marsh Lane, Preston. *Tel.:* Preston 54821.
    Enquiry Office: Windermere—*Tel.:* 3025.
Lake Windermere Steamers:
  Lakeside, Newby Bridge, Ulverston. *Tel.:* Newby Bridge 539.
Ullswater Navigation and Transit Co.
  13 Maude Street, Kendal.
  Tel.: Kendal 21626 or Glenridding 229.

RAMBLERS' ASSOCIATION

Lake District Area:
  Hon. Sec.: R. Taylor, 62 Loop Road North, Whitehaven.
  Tel.: Whitehaven 4719.

RESEARCH STATIONS

Freshwater Biological Association:
  Ferry House, Far Sawrey, Ambleside.
  Tel.: Windermere 2468.

ROYAL AUTOMOBILE CLUB

Headquarters:
  85 Pall Mall, London SW1. *Tel.:* (01) 930-4343.
Northern Counties Office:
  2 Granville Road, Jesmond Road, Newcastle upon Tyne 2.
  Tel.: Newcastle 814272.
    Patrol Service Centres: Keswick—*Tel.:* 73071.
                Bowness-on-Windermere—*Tel.:*
                          Windermere 3949.
    24-hour Service: Viaduct Car Park, Carlisle—*Tel.:* 26990.

TOURIST ORGANISATIONS

English Lakes Counties Tourist Board:
  Ellerthwaite, Windermere.
  Tel.: Windermere 4444. Telex 6592.

WATER SPORTS

Lake District Leisure Pursuits (for canoeing, sailing, water-skiing, rambling, etc.)*:*
  3 Crescent Road, Windermere. *Tel.:* Windermere 2296.
Royal Windermere Yacht Club:
  Secretary: A. Murdock; *Sailing Secretary:* W. B. Smith.
  Bowness-on-Windermere. *Tel.:* Windermere 3106.
Bassenthwaite Sailing Club:
  Hon. Sec.: M. Denwood, School House, Great Broughton, Cockermouth. *Tel.:* Brigham 436.

Y.M.C.A.

Headquarters:
  National Council of Y.M.C.A., 33 Endell Street, London WC2H 9AM.
  Tel.: (01) 836-3201.
North-West Regional Office:
  Div. Sec.: S. Childsworth, 83 Bridge Street, Manchester M3 2RF.
  Tel.: (061) 832-2156.
National Training Centre:
  Lakeside, Newby Bridge, Ulverston.
  Tel.: Newby Bridge 352.

Y.W.C.A.

Headquarters:
  2 Weymouth Street, London W1. *Tel.:* (01) 580-6011.
North-West Area Office:
  30 Ashley Road, Altrincham, Cheshire. *Tel.:* (061) 928-4478.
Hostel:
  Iveing Cottage, Old Lake Road, Ambleside. *Tel.:* Ambleside 2340.

YOUTH HOSTELS ASSOCIATION

Headquarters:
Trevelyan House, 8 St. Stephen's Hill, St. Albans, Herts.
  Tel.: St. Albans 55215.
Lakeland Regional Group:
  Church Street, Windermere.
  Tel.: Windermere 2301.

INDEX

Abbot Hall Art Gallery 84
Aira Force 100
Alston Moor 78
Ambleside 10, 55, 59, 70, 75, 84, 85, 88, 89, 92, 93, 94, 95, 103, 119, 125, 130, 131, 134, 135, 137
Angle Tarn 100, 105
Angling Crag 98
Appleby 85, 131, 134, 135
Appleby Castle 61
Arnold, Matthew 124
Ashness Bridge 100
Askham 55, 69
Aughertree Fell 57

Backbarrow 68, 74
Baddeley 89, 92
Bannerdale 57
Bannisdale 56
Banniside Moor 55, 57
Barley Bridge 75
Bassenthwaite 1, 47, 56, 57, 69, 70, 86, 90, 100, 126
Belle Grange 93
Bickness Combe 112
Birkrigg Oak Woods 24, 41
Birks Bridge 95
Black Beck 71, 75, 77
Black Combe 1, 28, 75
Black Sail Pass 98, 109
Blea Tarn 15, 89, 95
Blea Water 13, 15
Blencathra—see Saddleback
Bobbin Mills 75, 76
Boot 97
Boredale Hause 100
Borrans Field 56
Borrowdale 14, 24, 29, 55, 63, 65, 66, 67, 75, 77, 84, 89, 91, 92, 97, 98, 100, 101, 111, 112, 113, 130, 137
Borrowdale Volcanic Series 7, 10, 11, 12, 29, 53
Bowder Stone 84
Bow Fell 95, 96, 97, 105, 112
Bowness Hause 98
Bowness on Windermere 14, 62, 71, 92
Braddyll, Colonel 4
Braithwaite 105

Brandelhow 100
Brandreth 101
Broad Stand 108
Brockhole 84
Brontë, Charlotte 126
Brotherilkeld 97
Brotherswater 15, 112
Brough 85, 118, 122
Brougham 56, 88
Broughton in Furness 29, 65, 90
Brown, Dr. John 2, 3, 122
Budworth 122
Burn Moor 55
Burrow in Lonsdale 56
Buttermere 24, 59, 83, 84, 89, 90, 91, 98, 101, 105, 112, 113, 114

Caermote 56
Caldbeck 67, 77, 115, 122
Calder Abbey 59, 62, 84
Calder Hall 97
Calder, River 59
Carboniferous Limestone 8, 11, 22
Carlisle 3, 61, 69, 115, 119, 135
Carrock Fell 55, 126
Cartmel 60, 63, 66
Cartmel Fell 62
Cartmel Priory 62
Castle Crag 55
Castle How 57
Castlerigg 5, 53, 66, 89, 99, 125
Castle Rock of Triermain 111
Catbells 99, 100
Causey Pike 105
Chapel Stile 89, 94
Charcoal burning 68, 71, 72, 73, 74, 77
Claife Heights 92, 93
Clappersgate 88
Cleator Moor 73
Cockermouth 16, 61, 70, 84, 98, 123
Cocker, River 59
Cockley Beck 64, 88, 90, 95
Codale Tarn 134
Coleridge, Samuel Taylor 108, 124, 125
Colwith 68, 95
Comb Gill 112
Conishead 5
Coniston 10, 25, 32, 57, 63, 65, 67, 68, 70, 74, 77, 78, 84, 91, 94, 95, 96, 112, 116, 125, 126

Index

Coniston Limestone Group 11
Coniston Old Hall 73
Coniston Old Man 1, 13, 83, 92, 93, 94, 96, 126
Coniston Water 14, 15, 16, 47, 60, 68, 72
Copeland 60
Copper 67, 72, 77
Corney Fells 55
Cotton Mills 68, 75
Cowcove Beck 97
Cowmire Hall 68
Crinkle Crags 95, 96, 97, 105
Crosby Garrett 56
Crosby Ravensworth 56
Cross Fell 132, 133, 136, 138
Crosthwaite 65
Crummock Water 57, 64, 98, 101, 105
Cumberland 1, 3, 41
Cumberland County Council 47, 49
Cumbria 52, 58
Cunsey 68, 74

Dacre 55, 58, 61
Dalegarth Hall 97
Dale Head 55, 99
Dallam Towers 48
Deepdale 111, 112, 113
Deep Ghyll 109
Defoe, Daniel 122
Derwent, River 59, 75
Derwentwater 1, 2, 3, 5, 71, 78, 84, 86, 99, 100, 101, 111, 114, 122, 125
De Quincey, Thomas 123, 125
Devoke Water 97
Dickens, Charles 126
Dollywaggon Pike 101
Domesday Book 60
Dove Cottage 123, 125, 126
Dove Crag 105
Dovedale 111, 112
Dow Crag 94, 96, 112, 113
Duddon 23, 68, 73, 74, 88, 90, 91, 94, 96, 97
Dungeon Ghyll 33, 77, 89, 92, 94, 95, 105, 107
Dunmail Raise 15, 60, 70, 89, 93
Dunmallet 57

Eamont Bridge 70
Eden, River 37, 53, 57, 59, 61, 81
Eel Crag 105

Eel Tarn 97
Egremont Castle 61, 120
Ehenside Tarn 52
Elterwater 75, 76, 77, 93, 95
Elva 55
Ennerdale 24, 31, 32, 91, 98, 101, 103, 112, 114, 116, 131
Ennerdale Water 60
Esk, River 88, 96, 97
Eskdale 55, 56, 65, 73, 75, 81, 88, 90, 91, 95, 96, 97, 98, 114, 116, 130
Esk Hause 97
Eskmeals 20, 52, 53
Esthwaite Water 21, 34, 47, 93
Ewer (Ore) Gap 105

Fairfield 93, 103, 105, 131
Fell Foot Farm 88, 89, 95
Ferry 92
Fickle Steps 96
Fiennes 122
Finsthwaite Heights 76
Fleetwith Pike 98
Fletcher, Isaac 72
Flookburgh 51
Fox hunting 114, 115, 117
Friars Crag 5, 99
Froswick 105
Furness 60, 63, 66, 130, 133
Furness Abbey 62, 65, 72, 89, 122

Gatescarth Pass 105
Gilpin Bridge 135
Gilpin, William 3, 5, 6, 122
Gimmer Crag 107, 112
Glencoyne Park 56
Glenridding 86, 100, 101
Goats Water 13
Golden Eagle 39
Goldrill Beck 100
Gosforth 16, 53
Gosforth Cross 58, 60
Gowbarrow Park 100
Grange in Borrowdale 123
Grange over Sands 131, 132, 135
Granite 7, 67, 72
Grasmere 62, 67, 77, 83, 84, 85, 89, 90, 92, 93, 94, 101, 103, 114, 119, 120, 121, 122, 123, 124, 131
Grasmoor 105
Grass Guards 96
Gray, Thomas 4, 5, 6, 70, 122, 123

Index

Great End 86, 100, 130, 133
Great Gable 7, 12, 92, 97, 98, 100, 101, 105, 107, 109, 112
Great Langdale 14
Greenriggs 76
Greenside 77, 78
Greta Hall 124, 125
Grey Friar 95
Grisedale 56, 101
Grisedale Pike 99, 105
Grizedale Forest 37, 134
Gun powder 71, 75, 76
Gutterby Spa 19

Hall Dunnerdale 96
Hallin Fell 100
Hardknott Fort 56, 84
Hardknott Pass 88, 90, 95, 96, 97
Hart Crag 105
Harter Fell 95, 96, 97, 105
Haweswater 88, 105
Hawkshead 62, 63, 84, 93, 94, 123
Hawse End 100
Heathwaite 55
Heck Beck 57
Helm Crag 84
Helvellyn 1, 7, 13, 14, 31, 78, 84, 86, 91, 93, 94, 99, 100, 101, 102, 103, 126, 130, 131, 132, 133
Herdwick sheep 64, 128
Heron Pike 105
High Crag 83, 98
High Pike 105
High Street 13, 17, 88, 100, 105, 113, 114, 133
Hindscarth 99
Hobcarton Crag 105
Hodge Hill 68
Holme Cultram Abbey 62
Honister 10, 67, 78, 89, 98, 99, 101, 136
Horrax Mill 75
Hound trailing 114, 117
Howgill Fells 94
Howtown 100
Hugill 57

Ice Age 1, 12, 13, 14, 15, 16, 17, 133
Igneous rock 7,
Ill Bell 105
Ireby 57
Irish Sea 1
Iron 68, 72, 73, 74

Irthing, River 59
Irton Cross 58
Isel 75
Isel Church 58

Jack's Rake 107

Kendal 21, 60, 61, 62, 67, 68, 70, 74, 88, 89, 90, 130, 137
Kent, River 50, 74, 88, 131
Kentmere 57, 58, 64, 105
Kern Knotts 107
Keskadale Oakwoods 24, 41
Keswick 10, 55, 56, 59, 61, 63, 67, 68, 69, 70, 71, 83, 89, 90, 91, 99, 100, 103, 105, 113, 114, 122, 123, 124, 125, 131, 132, 133, 134, 135, 136, 137, 138
Kidsty Pike 105
Kilns 74
Kirkby Cartmel 62
Kirkby Kendal 62
Kirkby Lonsdale 62, 72
Kirkby Stephen 68
Kirk Fell 97, 98, 105
Kirkstone Pass 15, 39, 75, 84, 89, 101, 132

Lake District Planning Board iii, ix, 86, 93
Lamb, Charles 125
Langdale 15, 39, 71, 85, 89, 93, 94, 97, 105, 112, 113, 130
Langdale Pikes 53, 92, 93, 94, 95, 105, 107
Lanthwaite Green 57, 98
Latrigg 99, 127
Lead 72, 77, 78
Leighton 68
Levens 70, 135
Levens Estuary 50
Limestone 8
Lindale 68
Little Langdale 55, 56, 88, 89, 90, 95
Lodore 14, 99, 112, 125
Longsleddale 76
Loughrigg Terrace 93, 94
Loweswater 98, 107, 114
Low Fell 55
Low Water 13
Low Wood 68, 75, 77
Lyth Valley 85

Index

Maiden Moor 99
Manesty 100
Mardale 55, 56, 57, 65, 105, 114, 131
Martindale 23, 37, 55, 100
Maryport 16
Matterdale 132, 134
Mellbreak 107, 114
Mickledore 96, 108, 110
Millrigg 57
Mineral springs 70
Moor Divock 55
Morecambe Bay 4, 19, 20, 21, 50, 51, 59, 60, 61, 94, 130, 135
Muncaster 58, 75, 89, 97
Museum of Lakeland Life 84
Mungrisdale 122

Nab Scar 105
Napes Ridge 101, 112
National Nature Reserves 47, 49
National Trust 100, 127, 128
Newby Bridge 14, 70, 73
Newlands 24, 67, 68, 77, 84, 89, 90, 98, 99
New Red Sandstone 11, 12
Nibthwaite 14, 73, 74

Orrest Head 91, 92
Orton 68, 69
Osmotherley 60
Oxendale 105

Patterdale 85, 89, 100, 101, 113, 114, 119, 121
Pavey Ark 8, 105, 107, 112
Peel, John 115, 116, 120
Peel Wyke 57
Pennant, Thomas 2, 122
Pennines 28, 105, 110, 131, 132, 136, 137
Penny Bridge 68, 70
Penrith 15, 58, 90, 132, 135
Penrith, Old 56
Picturesque Beauty 2, 3, 4, 5, 6
Pike o' Blisco 95, 105
Pike o' Stickle 108
Pillar 97, 98, 103, 105, 108, 109, 112, 113
Place Fell 100
Plumbago 67
Pocklington, Mr. 5
Pooley Bridge 100, 122, 131

Portinscale 53, 100
Potter, Beatrix 84, 128

Queen Adelaides Hill 92

Rainfall 130
Rannerdale Knotts 98
Ransome, Arthur 128
Raven Crag 14, 103, 111
Ravenglass 19, 20, 47, 50, 56, 76, 88, 89, 97
Ravenglass Dunes 49
Ravenstonedale 68, 69
Red Bank 93, 94
Red deer 37, 80
Red Nab 93
Red Pike 97, 98, 105
Roe deer 43, 44
Rosthwaite 100, 131
Rothay 93
Rufus, William 60
Rush Bearing 85
Ruskin, John 84, 126, 127
Ruskins Brantwood 73
Rusland Moss 29, 41, 44
Rydal 84, 85, 90, 93, 104, 121, 123

Saddleback (Blencathra) 1, 14, 90, 99
St. Bees Head 49, 60, 131
St. Bees Sandstone 52
St. John's in the Vale 4, 55, 89, 90, 126
Sail Pike 105
Salmon 80
'Samson's Bratful' 53
Sandwich 100
Satterthwaite 74
Sawrey 84
Sca Fell 1, 7, 12, 15, 33, 53, 82, 88, 94, 95, 96, 97, 98, 99, 108, 109, 110, 111, 112
Scafell Pike 8, 53, 92, 96, 100, 103, 132
Scale Beck 98
Scale Force 84
Scales Tarn 14
Scarth Gap 98
Scoat Fell 98, 105
Scott, Walter 126
Screes 97, 98
Seathwaite 25, 81, 96, 112, 130, 131
Seatoller 89
Sedgwick, Adam 7
Shap 56, 67, 70, 132

Index

Shap Abbey 59, 62
Shap Wells 70
Sheep dog trials 85, 120, 121
Shelley, Percy 125
Shelter Crags 105
Shepherds Crag 111, 112
Silurian 8, 11, 12, 15
Silver How 93
Skelwith Bridge 89, 95
Skiddaw 1, 90, 93, 94, 99, 100, 101, 102, 103, 115, 123, 125, 132
Skiddaw Slates 8, 10, 11, 15, 28
Skirsgill Hill 57
Small Water 13
Smith Haskett 108, 109
Spark Bridge 68, 74, 76
Sphinx Ridge 107
Spooney Green Lane 99
Stake Pass 53, 105
Standing Stones 55
Stanley Ghyll 96
Staveley 59, 75, 76
Steadman, George 118, 119
Steeple 97, 98, 105
Stickle Tarn 77, 105, 107
Sticks 101
Stockdale Moor 53
Stoney Hazel 74
Stoney Tarn 97
Stoneythwaite Rake 96
Stott Park 75, 76
Striding Edge 13
Subberthwaite 65
Swart Beck 78
Swinside 55
Swirral Edge 13

Tarn Hows 43, 82, 94
The Three Shire Stone 84, 95
Thirlmere 14, 15, 37, 55, 56, 84, 90, 101, 111
Three Tarns 105
Threlkeld 72, 116
Thornthwaite Crag 85, 105
Thurstanwater 60
Thwaites Fell 55
Tilberthwaite Ghyll 33, 72, 78
Tongue House 57
Torver Beck 57
Troutbeck 55, 56, 88, 115
Troutbeck Bridge 76
Turnpike 70

Ullswater 15, 47, 48, 56, 61, 75, 77, 86, 88, 100, 125, 131
Ulpha 68, 75, 97
Ulverston 21, 64, 70, 72

Waberthwaite Cross 58
Waitby 56
Wallabarrow Crag 96
Walls Castle 56
Walna Scar 94, 95
Walney Island 49
Walpole, Hugh 84, 127
Wandope 105
Wasdale 91, 97, 98, 105, 107, 113, 130, 131, 132
Wasdale Head 65, 92, 97, 98, 101, 107
Wastwater 5, 14, 31, 73, 97, 101
Watendlath 84, 89, 99, 100
Watercrook 56
Waterhead 93
Wetherlam 92, 95, 113
Wet Sleddale 65
Whinlatter Pass 70
Whitbarrow 25, 26, 57
Whitehaven 2, 16, 131, 133
Whiteless Pike 105
White Moss Common 126
Whiteside 105
Whitrow Beck 57
Wilkinson, John 68
Wilkinson, Thomas 4, 69
Wilson, Professor Carus 63, 126
Windermere 3, 14, 15, 16, 59, 60, 62, 70, 71, 73, 83, 85, 86, 88, 90, 91, 92, 101, 114, 120, 121, 122, 125, 130, 134
Windermere Char 47, 80
Winster 68
Witherslack Spa 70, 75
Wool 74
Wordsworth, William 4, 7, 70, 84, 85, 91, 108, 123, 124, 125, 126, 129, 133
Workington 69
Woundale Common 55
Wray Castle 93
Wrestling 118
Wrynose Pass 56, 88, 90, 95
Wythburn 66

Yewbarrow 97, 98, 105
Yew Crag 100

The Country Code

GUARD AGAINST ALL RISK OF FIRE

Plantations, woodlands and heaths are highly inflammable: every year acres burn because of casually dropped matches, cigarette ends or pipe ash.

FASTEN ALL GATES

Even if your found them open. Animals can't be told to stay where they're put. A gate left open invites them to wander, a danger to themselves, to crops and to traffic.

KEEP DOGS UNDER PROPER CONTROL

Farmers have good reason to regard visiting dogs as pests; in the country a civilised town dog can become a savage. Keep your dog on a lead wherever there is livestock about, also on country roads.

KEEP TO THE PATHS ACROSS FARM LAND

Crops can be ruined by people's feet. Remember that grass is a valuable crop too, sometimes the only one on the farm. Flattened corn or hay is very difficult to harvest.

AVOID DAMAGING FENCES, HEDGES AND WALLS

They are expensive items in the farm's economy; repairs are costly and use scarce labour. Keep to recognised routes, using gates and stiles.

LEAVE NO LITTER

All litter is unsightly, and some is dangerous as well. Take litter home for disposal; in the country it costs a lot to collect it.

SAFEGUARD WATER SUPPLIES

Your chosen walk may well cross a catchment area for the water supply of millions. Avoid polluting it in any way. Never interfere with cattle troughs.

PROTECT WILD LIFE, WILD PLANTS AND TREES

Wild life is best observed, not collected. To pick or uproot flowers, carve trees and rocks, or disturb wild animals and birds, destroys other people's pleasure as well.

GO CAREFULLY ON COUNTRY ROADS

Country roads have special dangers: blind corners, high banks and hedges, slow-moving tractors and farm machinery or animals. Motorists should reduce their speed and take extra care; walkers should keep to the right, facing oncoming traffic.

RESPECT THE LIFE OF THE COUNTRYSIDE

Set a good example and try to fit in with the life and work of the countryside. This way good relations are preserved, and those who follow are not regarded as enemies.

OTHER ILLUSTRATED GUIDES IN THE NATIONAL PARK SERIES

Brecon Beacons	**50p** (61p)
Dartmoor	**45p** (54½p)
Exmoor	**50p** (61p)
Northumberland	**37½p** (48½p)
North York Moors	**45p** (56p)
Peak District	**85p** (94½p)
Pembrokeshire Coast	**75p** (88p)
Snowdonia	**75p** (88p)
Yorkshire Dales	**45p** (56p)

Prices in brackets include postage

Please send your orders or requests for free lists of titles of guide-books to Her Majesty's Stationery Office, PM1C Atlantic House, Holborn Viaduct, London EC1P 1BN

Government publications are obtainable from the Government Bookshops in London (post orders to P.O. Box 569, SE1 9NH), Edinburgh, Cardiff, Belfast, Manchester, Birmingham and Bristol, or through H.M.S.O. Agents and other booksellers

Printed in England for Her Majesty's Stationery Office by Headley Brothers Ltd, 109 Kingsway London WC2B 6PX and Ashford, Kent
Dd.504251 K88 4/75

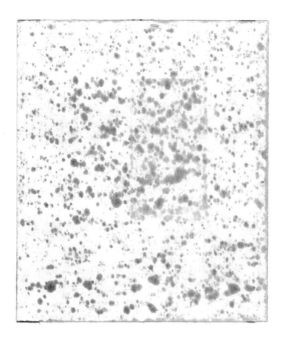